KUNDALINI AWAKENING

Ultimate Guide to Gain Enlightenment, Awaken Your Energetic Potential, Achieve Higher Consciousness, Activate and Decalcify Pineal Gland, Expand Mind Power, Intuition, Enhance Psychic Abilities, Divine Energy, Self-Realization

JENIFER WILLIAMS

KUNDALINI AWAKENING

© Copyright 2018 by Jenifer Williams - All rights reserved.

The following eBook is reproduced below with the goal of providing information that is as accurate and reliable as possible. Regardless, purchasing this eBook can be seen as consent to the fact that both the publisher and the author of this book are in no way experts on the topics discussed within and that any recommendations or suggestions that are made herein are for entertainment purposes only. Professionals should be consulted as needed prior to undertaking any of the action endorsed herein.

This declaration is deemed fair and valid by both the American Bar Association and the Committee of Publishers Association and is legally binding throughout the United States.

Furthermore, the transmission, duplication or reproduction of any of the following work including specific information will be considered an illegal act irrespective of if it is done electronically or in print. This extends to creating a secondary or tertiary copy of the work or a recorded copy and is only allowed with an expressed written consent from the Publisher. All additional right reserved.

The information in the following pages is broadly considered to be a truthful and accurate account of facts, and as such any inattention, use or misuse of the information in question by the reader will render any resulting actions solely under their purview. There are no scenarios in which the publisher or the original author of this work can be in any fashion deemed liable for any hardship or damages that may befall them after undertaking information described herein.

Additionally, the information in the following pages is intended only for informational purposes and should thus be thought of as universal. As befitting its nature, it is presented without assurance regarding its prolonged validity or interim quality. Trademarks that are mentioned are done without written consent and can in no way be considered an endorsement from the trademark holder.

Table of Contents

Introduction
Chapter 1: The Basics of Kundalini Awakening
Chapter 2: Chakras
Chapter 3: Prana
Chapter 4: The 4 Elements
Chapter 5: Akasha
Chapter 6: Meditation
Chapter 7: The Intuition
Chapter 8: Chakra Healing
Chapter 9: Psychic Abilities
Chapter 10: Astral Travel
Chapter 11: Connect to Your Higher-self
Chapter 12: 5-Minute Meditation Sessions
Chapter 13: 30-Minute Meditation Sessions
Chapter 14: Aura
Chapter 15: Vibration
Chapter 16: Mind Power
Chapter 17: Enlightenment
Chapter 18: Fasting

Chapter 19: Best Practices ....

Chapter 20: Road to Mastery ....

Conclusion ....

Introduction

Congratulations on downloading this book and thank you for doing so. *Kundalini Awakening: Ultimate Guide to Gain Enlightenment, Awaken Your Energetic Potential, Achieve Higher Consciousness, Activate and Decalcify Pineal Gland, Expand Mind Power, Intuition, Enhance Psychic Abilities, Divine Energy, Self-Realization* will you teach you to unlock the latent powers that you possess. Some of these practices have been in existence since ancient times and are now going to be revealed to you. As we enter into the Age of Aquarius, many people turn toward spirituality. This book will not only unleash your hidden potentials, but it will also help you turn into a spiritual being.

Let us look at what you will learn from the following chapters:

Chapter 1 is about the basics of kundalini awakening. Find out more about this ancient mystery and how you can tap into its amazing power.

Chapter 2 talks about the different chakras of the body. Chakras are energy centers that have many functions and attributes. They also ensure the free flow of energy in the body.

Chapter 3 is about the so-called prana. Everything in the universe is made of prana. Without prana, life cannot exist. Learn more about this amazing energy.

Chapter 4 discusses the world of the elements. Everything in creation is said to be composed of these elements. Learn about their attributes and fine qualities, as well as how you can use them to your advantage.

Chapter 5 talks about the akasha. Akasha is that mysterious force from which all the elements come from.

Chapter 6 is about meditation. When it comes to awakening the kundalini and enhancing your psychic powers, as well as in the search for enlightenment, the practice of meditation is the number one requirement that you should learn.

Chapter 7 teaches the secrets of the intuition. Learn how you can use and develop your intuition and connect to a higher power.

Chapter 8 is about chakra healing. Learn how you can heal your chakras naturally and through meditation.

Chapter 9 is about the amazing world of psionics. It discusses the different psychic abilities and the proper training that you can do to develop these abilities.

Chapter 10 is about astral travel. Learn how you can separate your spiritual body from your physical body and travel the whole universe.

Chapter 11 teaches how you can connect to your higher-self, also known as the God-self.

Chapter 12 gives 5-minute meditation sessions that you can do. These meditation techniques are excellent for beginners.

Chapter 13 teaches 30-minute meditation sessions. These meditation techniques are good for those who already have experience in the practice of meditation.

Chapter 14 talks about the aura. Learn how to sense and see auras.

Chapter 15 is about vibration. Learn how you can raise your vibration and be protected from psychic vampires.

Chapter 16 explores the power of the mind. Learn more about your mind and the infinite power that you possess.

Chapter 17 is a discussion on enlightenment. Learn about the meaning of true enlightenment and how you can achieve it in your life.

Chapter 18 is about the practice of fasting. Find out its importance and how you can benefit from it.

Chapter 19 lays down the best practices to help you succeed. Learn the correct attitude, mindset, and approach to learn the teachings in this book

Chapter 20 teaches the road to mastery from a beginner to adept.

May this book be your guiding light to joy, peace, happiness, success, freedom, and a meaningful life.

There are plenty of books on this subject on the market, thanks again for choosing this one! Every effort was made to ensure it is full of as much useful information as possible. Please enjoy!

Chapter 1: The Basics of Kundalini Awakening

The Nature of Kundalini

The kundalini is a primal force. It is also known as *Shakti* or *serpent power*. Some people refer to it as the goddess power. Regardless of the name by which you want to call it, it remains to be a force that can take your spirituality and psychic powers to the next level.

The kundalini is located at the base of the spine at the root chakra. It is called the serpent power because a dormant kundalini is coiled just like a serpent. However, once you awaken the kundalini, it shall uncoil itself and rise through the main chakras. Every chakra that it passes through shall be empowered and energized. Your kundalini shall rise up to the top-most chakra, which is the crown chakra. The crown chakra is located a few inches above your head. The awakening of the kundalini is closely associated with enlightenment and acquisition of psychic powers.

Kundalini vs. Prana

There are people who get confused between kundalini and prana. So, what is prana? Prana is energy. In spirituality, it is believed that prana exists everywhere. It is inside you and all around you. Nothing can ever exist without prana. Later in this book, you will learn how you can tap and harness prana. Mastery over prana will allow you to do wonderful things. This is because everything is made of prana. If you master the art of manipulating prana, then you shall be a master of everything. It should be noted that everything, including those that you do not see with your eyes, is made of energy (prana).

Take note that this energy is different from the energy that Einstein had discovered. This energy or prana is not limited to any formula. It is literally everything and pervaded everything. So, is the kundalini the same as prana? The answer is no. Prana is a broad term that encompasses all energy, while kundalini is more specific. Kundalini is that coiled-looking serpent that is at the base of your spine, waiting to be unleashed. It should also be noted that this serpent is not a gross material serpent but is ethereal – it is made of prana. Prana embraces everything. You are prana, and everything around you is also made of prana.

Kundalini vs. Chi

The answer here would be the same as the one given to kundalini vs. prana. This is because chi is just another term for prana. Chi is the Chinese term for prana.

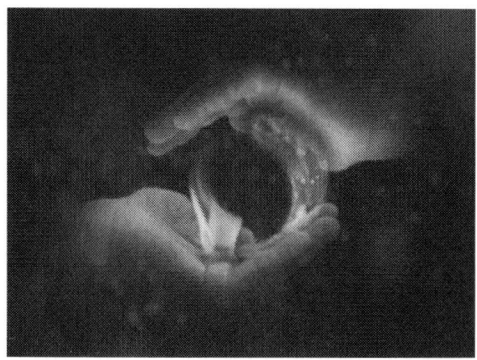

Prana has been in existence since the beginning of the universe. It has been called by many names, depending on the tradition and culture. In India, they normally refer to this energy as prana. In China, they call it chi. In Greece, they call it pneuma. In Japan, they refer to it as Ki. In Western countries, they often refer to it as energy. Still, all these terms refer to the same thing. Call it whatever you want. The important thing is that you understand what it is.

The Benefits and Effects of Kundalini Awakening

So, what happens when you finally awaken the kundalini? As already stated, it shall rise up through the main chakras up to the crown chakra. As it does, it shall energize and greatly empower all the other main chakras. Since it shall start from the base chakra, it will pass through all your main chakras up to the top. This is not just a simple passing of the kundalini, but there is also a great transference and awakening of power.

KUNDALINI AWAKENING

When the kundalini is awakened, and due to its tremendous primal force, there are certain sensations that you may feel. People who have awakened their kundalini reported an intense pressure and blissful feeling overflowing from within. It is pleasurable and is said to be more pleasurable than an orgasm. You can also have a feeling of having your mind expands to infinity. Also, as the kundalini rises, you can feel a great sense of bliss every time it touches your chakras.

Once the kundalini is awakened, you can also feel more sensitive, especially to subtle energy or prana. It will make you more empathic to everything. You will also have more control over yourself. In fact, people who have awakened their kundalini report a feeling of satisfaction, whereby they are no longer a slave to their desires and urges. There is also a feeling of peace and harmony.

Of course, a well-known effect of awakening the kundalini is the acquisition and development of psychic powers. Its awakening can also signify enlightenment.

There are many other effects of awakening the kundalini. Take note that there are no strict or hard rules on what you would feel. It will depend on your current spiritual maturity. What I have enumerated are the most common experiences when a person successfully awakens the kundalini power.

Clearing the Blockages That Prevent Kundalini from Rising Smoothly

The kundalini rises through the 7 main chakras starting from the root chakra at the base of the spinal cord. Hence, to ensure that it rises smoothly up to the top-most chakra, the crown chakra, you need to ensure that the channel from the root chakra up to the crown chakra is free from blockages. There are different ways on how to do this, but the best way is by meditation and to ensure that the chakras are strong and healthy. As you read through this book, you will learn how to purify and strengthen your chakras. The important key is to keep on practicing, especially the meditation exercises.

How to Awaken A Dormant Kundalini

There is no single rule or way to awaken a dormant Kundalini. It is believed that the kundalini will uncoil on its own once you have attained a certain level of spiritual maturity. Now, when it comes to growing or developing your spirituality, the best way to do this is through meditation. Later in this book, you will learn various meditation techniques, as well as many other practices. Although they might not directly work on the kundalini, they are the keys that will allow you to develop your spirituality, which will induce the awakening of the kundalini.

Now, there are also those who claim that the kundalini can be awakened by another person. This means that instead of doing the exercises and meditation techniques, you will simply let go of everything and submit yourself to a guru or master. Yes, it is possible for the kundalini to be awakened this way. However, it should be noted that this is not the recommended method because such awakening of the kundalini is only temporary. It is also good to give a student an idea of how it feels to have the kundalini awakened. The reason why it is only temporary is that the student's spirituality is not ready for it. Again, the real measure of the true awakening of the kundalini is one's spiritual maturity and development. Remember that this is not about gaining the power to advance your selfish interests. True kundalini awakening is about having the right mindset and spiritual readiness – only then will the universe grant you the gift of true kundalini awakening.

Chapter 2: Chakras

Understanding Chakras

When you deal with spiritual matters, you will definitely encounter the term, *chakras*. So, what are chakras? Chakras are energy centers in the body. The body has many chakras, major and minor. However, it is noteworthy that there are 7 major chakras, and they are located along the spine. Just as the physical body has its vital organs, the spiritual or astral body has these chakras. Every chakra has its own attributes and characteristics. As energy centers, they also ensure the free flow of energy throughout the body. It is important to keep your chakras strong, cleansed, and healthy. When a chakra becomes blocked or weak, then it will soon manifest as a physical ailment. Therefore, working on your chakras is not only for spiritual purposes but also for the good of your physical body.

The Chakra System

The chakra system is divided into seven main chakras. These are the main chakras of the body which are located along the spine. Let us discuss them one by one:

- Root chakra

The root chakra or base chakra is the first chakra in the system. It is located at the base of the spine. As such, it is also the place where the kundalini resides. Its color is red. The root chakra is the chakra of foundation and stability. This can easily be understood as the root chakra also supports all the other main chakras. The root chakra is associated with stability, support, foundation, being grounded and centered, physical identity, earth, safety, protection, and prosperity. If you have a weak root chakra, then you might have problems with having fear issues or anxiety. You can also lack confidence and feel weak. Another common issue would be self-doubt and a sense of helplessness. However, an over-activated root chakra can make you greedy of material things and be too focused on your personal gains. This is why it is important to balance this chakra to avoid adverse effects.

The root chakra also reminds you to be aware of the present moment, to be here and focus on the moment of now. Since this is also the place where the serpent power resides, it is extremely important that you ensure that you have a strong and healthy root chakra.

- Sacral chakra

The sacral chakra is also known as the sex chakra. It is located just below the belly button. It is the center of your sensuality and sexuality. This is the home of your sexual energy. Its color is orange. This does not refer to just sex alone but also your self-identity and expression. After all, one's sexuality governs their life and not just about romantic stuff.

If you have some issues with your sexuality or if you find it hard to express yourself, then you should work on this chakra. If you have an overactive sacral chakra, then you might have some troubles controlling your sensual nature. This is why you should keep this chakra balanced and under your control. This is also the next phase where the kundalini rises, so you need to make sure that this chakra is clear to make the kundalini rise freely.

- Solar plexus chakra

The next chakra is the solar plexus chakra. As the name implies, it is located in the solar plexus area. The color of this chakra is yellow. This is the center of your willpower and personal identity. It is also the center of your emotions. Be careful with this one as it can be hard to deal with negative and destructive emotions. In life, it is okay and normal to face challenges, so you need to have a strong will, and you should know your true identity.

If you feel like you fail to exercise your will or when things around you are not going the way that you would want them to be, then that is a signal that you should work on your solar plexus chakra. However, an over-activated solar plexus chakra can make you stubborn where you just do what you want without regard to other circumstances. It can cause greed and selfishness.

- Heart chakra

The heart chakra is the center of universal love and kindness. It is located in the center of the chest, and its color is green. This chakra is associated with relationships, emotional issues, and love. Take note that this does not refer to the romantic kind of love, but a universal kind of love. True spiritual masters have a strong heart chakra, and this is how they can love absolutely everyone. The heart chakra is also closely related to the crown chakra.

In fact, it is believed that the only way to fully activate the crown chakra is by first activating the heart chakra. If you want to be kinder and more loving to people, then you should strengthen your heart chakra. An over-activated heart chakra is not good because it might cause you to be abused by negative people. This does not mean that you should limit the power of your heart chakra, but be sure that you can control it. This is why working on any chakra requires that you also work with all the other main chakras. They all need to be developed at the same time. Do not worry, later in this book, you will learn just how you can enhance all the chakras at the same time to ensure that they are in harmony with one another and nothing is being left behind. The heart chakra is also very important to kundalini awakening.

- Throat chakra

The throat chakra is located in the throat, and its color is blue. This chakra is associated with communication and self-expression. If you find it hard to voice out your thoughts and ideas or difficult to communicate, then you should work on this chakra. There are also people who relate this chakra to creativity as it also governs self-expression. If overly activated, then this might cause a person to be too talkative and find it hard to keep secrets.

Just like all the other chakras, it is a must that you balance and keep your throat chakra under your control. Hence, overall spiritual growth and maturity are recommended. If you stutter, then that is often a sign that you should work on your throat chakra. Moreover, it should be noted that this chakra is not limited to oral communication, but it also governs written and all other forms of communication and expression of one's self.

- Ajna chakra

The Ajna chakra, or more commonly known as the third eye chakra, is the seat of the intuition. Now, it is important to take note of the correct position of this chakra. Specifically, it is located between the eyebrows, and its color is indigo. The psychic ability of clairvoyance (clear-seeing) pertains to this chakra. In fact, many psychic powers are connected to the Ajna chakra, since it connects you to the astral realm or the word of the spirits. It is also noteworthy that regular practice of any form of meditation strengthens the Ajna chakra.

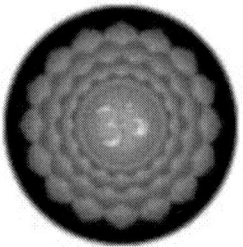

An important note has to be made: Many people think about how they can have a third eye or how they can open their third eye. Now, you should realize that everybody has a third eye. Yes, you have a third eye. Now, it is not really opening the third eye that matters but further developing it. The best way to do this is through meditation, especially meditation techniques that will require you to make use of visualization. Fortunately, such kinds of meditation techniques will be taught to you later in this book.

- Crown chakra

Last but not least, there is the crown chakra. The crown chakra is located a few inches above your head. Many people mistakenly think that it is located on the head itself, but that is not the precise location of the crown chakra. Remember that it is about two or three inches above your head. The crown chakra is the center of divine consciousness and enlightenment. It is also what will make you realize the oneness of spirituality. The color of this chakra is violet. It is the seat of illumination and divine oneness with all. This is the last chakra where the kundalini rises upon awakening.

The crown chakra is associated with wisdom, divine illumination and oneness, divine knowledge and understanding, and cosmic consciousness. You are probably aware of how leaders like kings and queens wear their golden crown. It is only a symbol of their enlightened crown chakra. Unfortunately, in the modern time, there are probably no enlightened leaders. The crown chakra is very important to enlightenment and full activation of the kundalini. Be sure to develop and strengthen your crown chakra.

A note about chakras

If you come to think about it, awakening the kundalini is not just about working on a chakra of your choice. To ensure that it fully rises through the chakras, you need to work on all the 7 main chakras. Now, do not worry about this as it is not as hard as you might think. You do not need to work on one chakra at a time, except if you notice that a particular chakra needs more healing or work. It is also noteworthy that all meditation techniques are like general exercises that strengthen and empower the whole chakra system. Be sure to keep your chakras clean and strong. They do not just ensure the good health of the physical body, but they are also necessary to the awakening of the kundalini, as well as to progress in your spiritual life.

Chapter 3: Prana

Understanding Prana

When you learn about spirituality, especially when it is related to the awakening of the kundalini and psychic powers, you will definitely encounter the term, prana. As we have already discussed everything is made of prana. Prana or energy, also known as life force or vital force, is inside you and all around you. Even modern-day scientists have already discovered that everything is made of energy. For years, it was believed that everything was just made of atoms and molecules. However, after looking deep into these molecules, there were atoms, and looking deep into atoms, the surprising fact was discovered: atoms are made of energy. Hence, everything is made of prana. Without prana, no can be no life. Hence, it is important for you to understand it, as well as how you can use it to your advantage.

When a person meditates, his whole system gets charged with prana. This has been experimented using Kirlian photography, a kind of photography that allows you to see a person's aura and energy body. This is why meditation is very important. This is not really something new. Even the great scientist, Nikola Tesla, said that to understand the universe, you should think in terms of energy. Be reminded that this is no longer limited to the formula given by Albert Einstein about gross material energy, but this also involves something that is spiritual.

The prana that is inside you is called personal prana/energy, while the prana that is outside of you is called universal prana. You can tap and use both sources if you want. Many advise that you should not always use your personal prana since it can be tiring once you have exhausted your storage of personal prana.

Therefore, if you know that you will be using lots of energy, it is good to get it from the universe.

Now, to better understand what prana is, it is best that you learn how to use and control it. This way, you can also feel it. With enough training with clairvoyance, you can even see this prana. Let us now move on with some practical exercises on prana so you can experience it firsthand.

How to Control Prana

One thing that you should understand is that prana or energy follows thought. Hence, a very effective way to direct and control prana is by using your mind. But, how exactly should you use your mind for this purpose? Well, the secret is to use the power of visualization. When people think of the word, visualization, they often think about the sense of sight. Although this is also true, you should know that to increase the power of your visualization, you should use as many senses as possible. Hence, if you visualize an elephant, for example, do not just see it, but also hear, smell, and feel it. When you work with prana, another sense that is very important is the sense of feeling. You should feel the energy as it flows and accumulates. In fact, you can control prana just by the use of feeling alone. This is called tactile visualization. Still, the best way is to use as many senses as possible.

Okay, so how exactly do you control prana? Well, here is an excellent exercise that you can do:

Relax and focus on your hand. Now, see and feel your personal energy or prana flowing inside you. Visualize that this prana flows and gathers in your hand. See and feel as the energy accumulates. Yes, this is all a visualization exercise, but this is also how you can effectively control prana. With enough practice, you will feel the energy as it gathers in your hand. Remember to also put emphasis on your sense of feeling as you practice visualization.

This is actually easier than it may seem. Now, remember to focus on what you are doing. Do not think and say, "Oh, this is just my mind playing tricks on me." The time for you to do that will be *after* the exercise. Practice this regularly, and you will be able to control prana.

Here is another exercise that you can try:

Rub your hands together. Now, position them in front of you as if you were holding a small ball. Visualize drawing prana from the universe and gather it into a ball between your hands. Prana can take any form so you can visualize it in any way you want. For beginners, it is recommended to just see this prana as white light. See and feel as it flows and gathers into a ball between your hands. Do you feel it? Focus on it and let your visualization take reign. Feel and be open to sensations. Relax and let the energy flow.

Once you learn these 2 basic exercises, you will realize just how easy it is to control and direct prana. Later in this book, you will learn meditation techniques that will teach you how to use this skill to your benefit. For now, focus on the theories and the basics.

The Nature of Prana

Let us now talk about the nature. Prana is infinite, and it cannot be destroyed. Yes, prana does not die. Instead, it can only be altered from state to state. This is why it can be said that everything is eternal. Prana is also not limited to space and time. It is boundless and endless. Hence, there are people who think of prana as God. Prana is everywhere, and those who know how to control it can harness its great power. However, to successfully control prana, you need to have a well-trained mind. This is another reason why you need to practice meditation since it trains you on how you can use the power if your mind effectively. This is because prana is too sensitive. If you lose your focus, then it can quickly dissipate and go somewhere else. Therefore, you have to control your mind. Once you can control your mind, then you can also control prana. Do not worry, if you stick to the practices in this book, then you will learn how to effectively control your mind and make good use of prana.

Chapter 4: The 4 Elements

The Elements

Everything is made of the 4 elements. In the universe, there are 4 mighty elements: fire, water, air, and earth. It should be noted that these do not just refer to the gross material substance, but there are also certain spiritual and characteristics or qualities that are ascribed to them. Knowledge of these qualities will give you the key on how to successfully take advantage of these elements to your own benefit.

Fire

The element of fire is said to be the very first element that existed in creation. It is associated with willpower, lust, sex, pleasure, courage, bravery, war, determination, and others. Of course, it also related to the material fire as you know it but do not forget its qualities. If you feel like being afraid or losing hope, then you can find strength in the element of fire. Other associations of the fire element include salamanders, south direction, rosemary, cactus, masculine, noon, Hestia, firedrakes, clairvoyance, and others. If you want to develop your third eye, then you should work with the fire element as it brings light to everything and nothing can be hidden from it.

Water

The next element is water. It is associated with healing, relaxation, beauty, cleansing, emotions, and others. Its direction is the west. If you want to do some healing or want to be more beautiful, then you should stay close to the water element. Other associations of the water element include the Goddess Aphrodite, the color blue, inverted triangle, sense of taste, undines, nymphs, cup and cauldron, mirror, and silver, among others. Its animals are the fish, dragons, water snakes, swan, and crab. When you work with cleansing or healing, you always make use of the power of this element.

Air

The next is the element of air. The element of air comes from the east. It is associated with communication, intellect, gossips, dawn, smell, and hearing, knowledge and wisdom, spring, wand, and sword. Its animals include the raven, spider, and the eagle. It has a masculine gender and is yellow in color. If you want to work on your intellect or if you want to be a free spirit like a gypsy, then work with the air element.

Earth

Of course, there is the element, earth. This element is the one that is closest to humanity. No matter what a person does, they will always be one and connected to this element as each one of us swims in the energy of the earth. The element earth is also associated with the following attributes: grounding and centering, prosperity, materialization, fortune, stability, and foundation. It has a feminine gender, and it governs the north direction. It is also associated with the following gods and goddesses: Ceres, Athos, Adonis, Gaea, Rhea, Mah, Arawn, Pan, and Dionysus. Its elemental beings are the gnomes and trolls. Its color is green or brown. Being earth dwellers, it is only right that you get attuned to the earth element.

Elemental Application

Knowing the attributes and qualities of the elements is one thing; being able to use them to your advantage is another. Now, it should be noted that there is no hard and fast rule as to the correct attributes or qualities of the elements as they may differ to an individual. For example, a person who is more attuned to the fire element might find fire more relaxing than water. Hence, feel free to get to know and connect with these elements, and see how they work for you. Still, there are general meanings, symbols, or attributes that you can rely on, like water being the element of cleansing.

So, how do you make proper use of the elements? Well, once you know how each element affects you, you can use them to your advantage. For example, if you are feeling stressed, you might want to use the element water to remove all that stress. A good way to do this is by soaking yourself in the tub for a longer period or going to the sea. If you want to develop willpower, then you should go to the element fire. You can easily do this by sunbathing while visualizing that the sun's rays are empowering your will. If you want to be more stable, then be connected to the earth. Feel the earth with your feet and be more connected to it. If you feel like you are having a hard time to move on from something, then learn from the element air and be free. It is not really hard to learn from the elements and to apply them in your life. You just have to get to know them better, and how they affect you, so you will know the best way to use them. There are, of course, many variations and applications of this knowledge. It is not good to teach everything as you will have no opportunity to make your own reflections and discoveries. Mastery of this subject can allow you to do wonders and definitely change your life for the better.

Chapter 5: Akasha

Understanding the Source, Akasha

What is the Akasha? Akasha is more than an element. Rather, it is believed that all the four elements come from Akasha. It is the source of all, the origin of everything. There are even practitioners who consider akasha to be the god principle. It is worth noting that Akasha is not considered an element, yet it possesses all the elements. It is associated with the colors white and black. It is timeless and spaceless. It is the beginning and the end. It is infinite. Indeed, the way most people describe God can be attributed to the qualities of the akasha.

Being in possession of all the colors and all the qualities and elements, mastery of the akasha will allow you to master all the elements in the universe. However, be noted that this is not as easy as it may seem. To be able to master it requires a high level of spiritual maturity and development. Nonetheless, this is something that you can do as you progress on your spiritual journey.

Just as the elements come from the akasha, this means that everything in the universe, whether visible or invisible, has the energy of akasha. Just as Akasha is everywhere, nothing can ever escape its power.

It is also believed that the akasha contains the records of everything, which includes the past, present, and even the future. If you can develop a clairvoyant ability, you can look into the Akashic records and be able to foretell the future. This is the explanation of how many prophets and diviners can tell the future.

The realm of the akasha is the astral plane. This is why the astral plane is also timeless and spaceless. It is noteworthy that every physical thing has an astral counterpart. In fact, before a physical thing manifests, it will first have an existence in the astral plane. It should also be noted that all planes are one. They are just different in terms of vibration. Needless to say, among all of the elements, akasha is the one with the highest vibration. You do not really need to master akasha just to benefit from it, for such mastery might take years or even a lifetime to achieve. However, as you work with akasha, you will notice how your chakras and psychic abilities improve, as well as your overall energy system.

Later on in this book, you will learn a meditation technique that will teach you how to master the akasha and use it to develop certain psychic abilities. In fact, there are practitioners who focus only on mastering the akasha for it is the key to everything. However, it should be noted that gaining supernatural powers should not be the focus of one's spirituality. The acquisition of psychic abilities is only a part of the awakening of the kundalini. Remember to keep your focus on your spirituality and not be blinded by power.

Chapter 6: Meditation

What Is Meditation?

Regular practice of meditation is very important to a spiritual life, as well as in the awakening of the kundalini. But, what is meditation? When people think about it, they often imagine monks who spend hours with their eyes closed and body still, chanting mantras and experiencing strange things. There are even people who think of it as some form of satanic practice. So, what exactly is meditation? Meditation is the way to train your mind, release stress, still the mind, improve your focus and concentration, and also a form of prayer. As you can see, there are many definitions that you can give to meditation. This is because the true meaning of meditation is what you give it. If you only see it as a way to release stress, then that is meditation as far as you are concerned. However, it is also worth noting that meditation can also be used to awaken the kundalini, enhance your psychic abilities, and also to achieve enlightenment. It is now up to you how much you want meditation to be a part of your life.

It is also important to note that meditation is not solely a satanic practice. In fact, the practice of meditation is present in different religions and spiritual practices. It does not belong to any group or religion. Simply put, the practice of meditation is open for everyone.

It is hard, if not impossible, to understand what meditation is without experiencing it firsthand. The only way to learn what meditation is really all about is to practice it regularly. Let us now move on to the practical steps.

Important Guidelines

Before we proceed to the meditation proper, you should keep in mind some important guidelines to observe how to meditate properly. Let us discuss them one by one:

- Proper posture

When you meditate, it is important that you observe the right posture. Now, there are many ways to meditate. You can meditate while lying, sitting, standing, or even while walking. However, there are certain pros and cons for every posture. When you meditate in a lying down position, you can more easily relax. However, a common problem with this posture is that you can also easily fall asleep, which is a problem for many beginners. When you meditate in a standing position, although it will prevent you from falling asleep, it will be harder for you to focus since you will have to exert some physical effort to hold the position. Not to mention, this posture can be tiring after some time. The same reasons apply when you meditate while walking or doing something else. Now, as for the most recommended posture for meditation, it is the sitting position. This posture enjoys the benefits of being awake and focused and also not falling asleep. This is also the most recommended posture by many spiritual masters. In fact, even the great Buddha achieved enlightenment while meditating in a sitting position. You can meditate while sitting on a chair or even on the floor. If meditating on the floor or on your bed, you might want to put a pillow below your tailbone to make you feel more comfortable.

- Keep your spine straight

Regardless of the meditation posture that you use, you should always keep your spine straight. This is very important when you meditate. The reason why you keep the spine straight is to ensure the free flow of energy through the seven main chakras. Remember that your main chakras are located along your spinal cord. By keeping it straight, the energy flows freely and smoothly, which is also important in awakening the kundalini. Now, a common mistake is to slouch during meditation. Before you even attempt to meditate, be sure to teach yourself how to assume the proper posture.

In the beginning, you might find it a bit difficult not to slouch especially when you are in a sitting position. Do not be discouraged. With enough practice, your body will learn to adjust, and you will get good at it. Therefore, just keep on practicing.

- Focus

When you meditate, you will be asked to focus on something. It can be a mere visualization exercise or a simple sound or object. Do not allow your mind to wander. Remember what your point of focus is in a certain meditation practice and stick to it. If there are other thoughts that arise in your mind, simply ignore them and bring your focus back to the object of the meditation.

Now, the most common challenge when a person learns to meditate is known as the monkey mind. What is the monkey mind? It is a state of mind that is full of thoughts. Just like a monkey that jumps from one branch to another, so does the mind leap from one thought to another. This is what is called as monkey mind. People who are just beginning to meditate will definitely face this challenge. So, how do you deal with the monkey mind? There is only one way to overcome the monkey mind, and that is to continue practicing meditation. The more that you meditate, the more that you will be able to control and still your mind. There are no shortcuts except continuous practice.

- Relax

When you meditate, you should be as relaxed as possible. Allow your body to fall asleep and let your mind become free. Do not think that what you are doing is difficult. This will only put unnecessary strain and pressure on yourself which will prevent you from reaching a higher state of consciousness. Just relax and place all your focus on your meditation practice. The more that you put pressure on yourself the more that you will not be able to free yourself from your physical body. Relax so you can become light. When you are light enough, then you will transcend into a higher state of consciousness. The key is to relax and let go.

- You are safe.

Perhaps one thing that you should know is that the practice of meditation is safe. Unfortunately, there are some people who think that meditation can kill you or put you in danger. Well, as long as you do not meditate while driving or while crossing the street and similar circumstances, then rest assured that you are safe. In fact, according to experience to meditators and gurus, if ever your physical body is put in danger while you are in a state of meditation, such as if a fire breaks when you meditate, you will be instantly brought back to your physical body. Therefore, when you meditate, do not worry about your safety. This will only cause your energy and attention to be divided. Instead, put all your focus on your meditation practice.

It is recommended to only teach you the important guidelines when meditating, since having too much information can tempt your mind wander even more. It is now time for you to learn and finally experience what medication is truly all about. Remember that even if you do not notice any changes, even in your state of mind, do not be discouraged. With regular practice, you will get better and better at doing meditation. Continuous practice is the key. Let us now move on to the meditation proper.

Basic Meditation Practice

For your first meditation, it is good to learn about the meditation on the breath, or simply breathing meditation. This is probably the most basic meditation technique in the world. However, do not underestimate the power of this meditation. Many spiritual masters and even the great Buddha practiced this meditation for a long time. It is also not uncommon even for experienced monks to stick to this meditation for years. The power of this meditation lies in its simplicity. The steps are as follows:

Assume a meditative posture. Just relax and do not think about anything. Now, focus on your breathing. Breathe in, and out gently. In.... Out... do you realize just how sensitive life is? This in and out breathing cycle simply cannot be interrupted, or it would mean danger or even death. Appreciate the beauty of your breath. Breathe in and breathe out. Behold the gift of life. Now, put all of your focus on your breathing. Do not think about anything else, and just be mindful of your breathing. Relax and let go. Be one with your breath.

You can do this meditation technique for as long as you want. If you are just starting out, you might want to do it for a few minutes. However, it is also noteworthy that many experienced meditators do this medication for an extended period, even for hours. For now, just do whatever is comfortable for you. After all, it is not good to rush your spiritual progress. Just be sure to always do your best and be committed to it. It is also worth noting that this simple meditation energizes and empowers all your chakras. Regular practice of this meditation will allow you to reach such a state of mind that you have never experienced before. By the time you finish doing this meditation, you will be in a peaceful and harmonious mindset.

Common Pitfalls

Let us now talk about the common pitfalls or mistakes in meditation. It is good for you to be aware of these pitfalls so that you can avoid committing them. Still, it should be noted that many of these pitfalls are quite difficult to avoid, so do not be discouraged if you fall into them even though you have already been warned. Just continue to do your best and keep on practicing.

- Thinking

You should understand that meditation is not about thinking, but it is more about doing. It is not about thinking of the present moment but about being in the present moment. When you meditate, do not think, just be.

Another common mistake is to think, "Am I doing it right?" while you are meditating. If you think about this, then you are doing it the wrong way. The time to ask yourself this kind of question should be after the meditation but not during the meditation proper itself. When you meditate, you should not allow yourself to be divided in any way.

- Wrong focus

It is true that when you meditate, you usually have to focus on something. Unfortunately, some meditators end up focusing on focusing instead of actually focusing on the point of an object in meditation. Okay, this might be a bit confusing, but it is very important for you to understand this lesson. When you meditate, do not tell yourself, do not tell yourself that you need to focus on this or that. Instead, you have to do it. To focus entails an actual action and not a mere command. There is a difference between actually focusing on your breathing and telling yourself that you should focus on your breathing. You should give yourself some time to reflect on this, and be sure that you understand it completely.

- Scratching

When you meditate, you may sometimes feel like a part of your body may get itchy. Of course, the tendency would be to scratch it. The problem with this is that the mind will return its focus to the physical body. If you keep on doing this, then you will not be able to reach a trance state or a deeper state of consciousness. So, what should you do? Well, you just have to ignore it. Although it is normal to feel itchy while meditating, you must not let it bother or distract you. This might be quite difficult to do in the beginning, but you simply have to resist it and get used to it. With enough practice, you will no longer be bothered by this itchy feeling. In fact, you will not even think about it. Remember to just ignore it. The more that you think about it, the more distracting it will be. Once again, continuous practice is the key.

- Not enough practice

Learning to meditate is just like learning a new skill. And, just like learning a new skill, it requires practice. Do not expect that you can do well if you just practice once a week. If you are serious about your spiritual development, you should make it a priority to meditate at least once every day. If you are just starting out, you can just do even just five or ten minutes of meditation per day. However, as you improve and get used to it, you have to practice more. Continuous practice is important. It is suggested that you set up a schedule when you will meditate and stick to that routine. This way, you can be sure that you make time for your meditation practice. Without continuous practice, then it is impossible for you to reach a good level of spiritual maturity.

- Falling asleep

This is a very common mistake committed by beginners. The good news is that this can be prevented. Now, there a few points for you to remember. First, if falling asleep is an issue, then you should not meditate in a lying down position as it is the posture that can easily make you fall asleep.

You might also want to meditate somewhere else other than your bed. The bed is usually a message to the mind for taking a rest or for sleeping. Hence, when you meditate on your bed, it can signal the brain that it is time for you to sleep instead of engaging in meditation. Again, the recommended posture is the sitting position. Second, you should not meditate when you are tired. Needless to say, it is easier to fall asleep when you are tired. Third, you should not meditate before bedtime. Many beginners who meditate before they sleep at night end up falling asleep completely. This is because by around this time, you would already be sleepy and exhausted. Instead, what you can do is to meditate early in the morning. Of course, if you do not have issues with falling asleep while meditating, then can skip these preventive measures. In that case, feel free to meditate even in the evening. The important thing is to find out what works best for you.

But, do not be too hard on yourself. If you do fall asleep despite observing the said preventive measures, it probably means only that you needed the rest, so just try again next time.

- Having expectations

When you meditate, you must place all of your focus on the object of your meditation. This also means that you must let go of all of your expectations. Having expectations, whether something like this or that would happen, will only divide your energy. Having expectations will also prevent you from giving your 100% focus. Instead of having expectations, just relax and stick to your meditation practice.

- Being too hard on yourself

Do not be too hard on yourself. Even if you think that you are not able to meditate properly, just take it easy on yourself. Being frustrated is only counterproductive to your efforts. Instead of being hard on yourself, you should relax and think about the matter carefully and gently. Find out the mistakes that you have been committing, and make the right adjustments to correct them. You will be able to meditate more effectively if you are feeling happy or when you are filled with positive energy. Instead of imposing pressure on yourself, which is not even good for you, you should use that energy to focus more on your practice and do your best. Take note that even if you do your best, do not expect to be enlightened or awaken your kundalini right away. This is something that really takes lots of practice, time, and efforts. Instead, you should appreciate that you are on the right track and that you are well on your way to mastery.

Chapter 7: The Intuition

The Third Eye

When it comes to developing the intuition, the number one thing that would be discussed would be the third eye. The third eye is the seat of the intuition. It is the key to the power of clairvoyance, also known as *clear seeing*. The third eye or Ajna chakra is probably the most common chakra that many people are familiar with. It is what will allow you to see into the world of spirits. Remember that the exact location of the third eye is right between your eyebrows.

The good news is that everyone has a third eye. It is just a matter of developing it, and this is something that you can do. Once you develop your third eye, you will have a powerful intuition, and you will even be able to access the Akashic records. It will depend on how you make use of it. Another interesting reason to develop the third eye is to be able to see prana more clearly. There are many things that can be associated with the third eye, but the most common of all is the intuition.

Every person has some level of intuition. For example, have you experienced simply knowing who is calling your phone even without looking at it? This is a classic example of the use of the intuition. Of course, there are many other practical uses, such as being able to avoid danger or simply knowing the right course of action to take in a difficult situation. Indeed, developing the intuition can be very helpful. Let us now look more into enhancing this natural and psychic ability.

The third eye is also the pineal gland. It is a small endocrine system that regulates the wake-sleep pattern. In spirituality, when you talk about the pineal gland, then you also refer to the third eye.

Activate and Decalcify Your Pineal Gland

The pineal gland or your third eye holds remarkable power. However, only a few people can tap into this power and use it effectively. Many people simply have an underdeveloped third eye. But, the good news is that there are exercises that you can do to strengthen your third eye so you can start using and enjoying its immense power. Let us discuss them one by one:

- Who is it?

This is something that you can do every time your phone rings or beeps. Simply ask yourself, "Who is it?" Pay attention to what you see in your mind's eye. Do you see any image or impressions? Be open to receiving messages. This is how you can connect to your intuition. You should also realize that you have a strong intuition and that you only have to learn to connect to it. Of course, this technique is not limited to calls or texts on your phone. You can also adjust it a bit and use it in other ways. For example, if you hear a knock or any sound at night, you can ask, "What is it?" and pay attention to any messages that you get from your intuition. The important thing is to start connecting to your intuition once again.

- Forehead press

This technique is becoming popular these days. This, however, does not work on everyone but it is still worth trying. This will allow you some specks of prana in the air. They usually appear as little dots or any form of white light. The steps are as follows:

Place your index finger in the area between the eyebrows where the Ajna chakra is. Press it gently and maintain pressure for about 50 seconds. Slowly remove your finger, blink your eyes around five times, and look at a blank wall. Just focus lightly and try to see with your peripheral vision. Do you see little dots or any specks of white light? This is prana in the air.

To help you see the energy, you might want to do this in a dimly lit room. Look at a wall with a neutral background. This is a good way to use your third eye to see energy, but it is not a recommended method to strengthen the Ajna chakra. Still, this is something that is worth trying, especially if you just want to see prana.

- Visual screen

This is a good technique to use for visualization exercises. To locate the visual screen, close your eyes and look slightly upward. With eyes closed, look at the area of the Ajna chakra. This is your visual screen. You can project anything that you like to this screen, especially images. You can consider this as some form of internal magic mirror.

The main purpose of this visual screen is for your visualization exercises. Here is a simple exercise you can do to increase your concentration and willpower:

Assume a meditative posture and relax. Now, look at your visual screen. Imagine an apple floating in front of you. Now, just focus on this apple and do not entertain any other thoughts. This is just like the breathing meditation. However, instead of focusing on your breath, focus on the apple in your visual screen.

When you are ready to end this meditation, simply visualized the apple slowly fade away and gently open your eyes.

You are also welcome to use any other object for this meditation. If you do not want to use an apple, you can visualize an orange or even an elephant. The important thing is to have a point of visual focus for this meditation.

- Charging with the fire element

Remember that the intuition is associated with the pineal gland, in the pineal gland is the third eye chakra. Now, this third eye chakra is associated with the fire element. Therefore, you can empower your third eye chakra by charging it with the element of fire. This is a powerful technique so be sure to use it carefully. The steps are as follows:

Assume a meditative posture and relax. Close your eyes. Now, visualize the brilliant and powerful sun above you. This powerful sun is full of the element of fire. As you inhale, see and feel that you are drawing the energy from the sun. Let the energy charge your third eye chakra and empower it. Do this with every inhalation. The more that you charge your third eye, the more that it lights up and become more powerful. Have faith that with every inhalation, you become more and more intuitive.

Keep in mind that this is a powerful technique. If you are just starting out, it is suggested that you only do up to 10 inhalations in the beginning. You can then add one or two more inhalations every week. You will know if you can execute this technique properly because you will feel pressure on your forehead in the area of your third eye chakra. Take note that you should not just visualize your third eye chakra getting stronger, but you should also be conscious that your intuition becomes more powerful the more that you charge your third eye. The power of visualization should be accompanied by your intention.

Note

It should be noted that the Ajna chakra and crown chakra are closely connected. If you want to improve your intuition, it is only right that you also work on your crown chakra. Of course, this does not mean that you should ignore your other chakras. Again, the whole chakra system is important to your spiritual development and to the awakening of the kundalini.

Chapter 8: Chakra Healing

Natural Ways to Heal Your Chakras

Since the whole chakra system is important to your spiritual growth, it is necessary that you learn how to heal your chakras. Take note that the chakras are closely connected to your physical body. This is why if there is a problem with your chakras, it will soon manifest in your physical body. In the same way, if you keep your chakras clean and healthy, then your physical body will also be in good health. This also reveals how you can keep your chakras healthy. The simplest way to do it, especially for those who are not into meditation and similar practices is simply to keep your physical body fit and healthy. Hence, regular exercise is important. In fact, exercise is the most natural way that you cleanse and empower your chakras. It is not just about removing stress, but it also removes negative energies in your system. Now, you do not really have to go to the gym or lift some weights to do this; doing light exercises such as jogging or walking will also be fine. The important thing is to love that body and shake off those negative energies.

Another thing that you want to do is to eat healthily. Stick to eating vegetables and avoid processed foods. Many also advice to avoid eating meat and dairy products. As much as possible, choose greens and other healthy and truly nutritious foods. Now, it is hard to eat healthy right away. If you want, you can start making small changes to your diet. Do it gradually.

You should also make positive changes to your lifestyle. If you are a heavy smoker, then it is time for you to stop or start decreasing the number of sticks that you smoke per day. If you are a heavy drinker, then it is time for you to start drinking moderately. The more that you keep your physical body healthy, the more that you can cleanse and energized your chakra system.

However, let us admit the truth that it is simply hard to be healthy especially when you are used to so many unhealthy ways. In this case, it is unfortunate how you do not realize how much you damage your system by your unhealthy choices. However, things do not have to be like this forever. Be happy for the fact that you have experienced what it means to be unhealthy (in case you are unhealthy), but it is now finally time for you to start thinking about what is right for your body, physical and spiritual. In fact, in some traditions, it is forbidden to drink liquor and even though eat meat when you aspire to awaken the kundalini. Instead, you should strive to be healthy in all aspects of your life. The good news is that you do not really have to impose such strict discipline unless you are a follower of such strict and extreme traditions. However, it is important that you learn to be healthy. Needless to say, it is not just knowing to be healthy that matter, but it is also important that you live a healthy life. Once again, you need to take positive actions and start making healthy choices. This can only be hard in the beginning, but you will soon get used to it. In fact, you will soon realize how good it is to be healthy that you might find it hard to believe how unhealthy you have been.

Chakra Healing Meditation

One of the best ways to keep your chakras clean and healthy is by doing meditation. Although all meditation techniques heal the chakras, there are certain techniques that primarily focus on cleansing and energizing your chakras. It should be noted that although keeping your physical body healthy can be very helpful, it is still not enough to make your chakras fully cleansed and energized. The following meditation technique will keep your chakras cleansed and energized. You can do this meditation every day or whenever you feel necessary. If you feel like you have been exposed to negativity, then this is a good way to cleanse yourself and feel revitalized. The steps are as follows:

Assume a meditative posture and relax. In this meditation, you will charge your every chakra with the corresponding color. You will start with your root chakra and up to your crown chakra. Just relax, and breathe in and out gently.

Now, see and feel your base chakra. Remember that your base/root chakra is the chakra of foundation. Its color is red. Visualize a red ray of light descending from heaven. Allow this ray of light to enter the top of your head, and see and feel as it charges your base chakra. As you do, feel how the qualities of this chakra get empowered. You should also visualize your root chakra getting cleansed and energized. The more energy that you put into it, the more powerful it becomes. Keep charging it for as long as you like or until you feel strongly grounded and centered. When you are ready to move on to the next chakra, simply visualize the red ray of light turn into an orange color. This is a signal that it is now time to move to the next chakra, the sacral or sex chakra. Remember that the sacral chakra is located below your navel right in the public area, and its color is orange. Allow the orange ray of light to cleanse, energize and fill your whole sacral chakra. As you do, feel the qualities and attributes of this chakra being empowered. Do not hurry, and do not stop until you are satisfied with the energy that you have absorbed. However, you do not need to absorb too much energy. If you are just starting out, then just charge yourself as long as you feel comfortable with it. When you are done working on your sacral chakra, it is time to go to the next chakra, which is the solar plexus chakra. The solar plexus chakra is the center of your personal strength and willpower. Its color is yellow. Now, visualize the ray of light turn into a yellow color and allow it to feed your solar plexus chakra with lots of energy. Feel your personal and willpower getting stronger and stronger. Keep charging your solar plexus chakra. When you are satisfied, you can now move to the next chakra, which is the heart chakra. Visualize the ray of light turns into a green color. Now, let it charge and cleanse your heart chakra. Feel how your heart chakra lights up and is full of divine love for all sentient beings. Next, you will now charge the throat chakra. It is the chakra of communication and self-expression, and its color is blue. Let the ray of light turn into a blue color and allow it to charge your throat chakra. See and feel how your throat chakra lights up and is energized. Spend as much time as you want to charge your throat chakra. Now, let the ray of light turn into an indigo color and let it charge your Ajna chakra. This is your third eye, and it is the seat of your intuition and the chakra of clairvoyance. Allow the indigo ray of light to charge your third eye. Feel it being cleansed and fully energized. Finally, visualize the ray of light turns into violet. It is ow time to charge and cleanse your crown chakra. The crown chakra is located a few inches right above your head, and its color is violet. It is the center of higher understanding, divine consciousness, and higher wisdom. Continue to charge your crown chakra with divine energy until it is fully cleansed and energized. Now, see the ray of light slowly fade away. Visualize all of your 7 main chakras at the same time, and see and feel how they are all cleaned and energized. See how they shine brightly. You are feeling cleansed, empowered, and happy. Slowly bring your attention to your physical body. Move your fingers and toes, and gently open your eyes.

Feel free to spend as much time as you need with each chakra. Do not stop until you feel that the chakra that is being charged is fully cleansed and energized. You should also remember the attributes of the chakra, and be fully convinced that qualities or attributes are being empowered as you charge the chakra with energy. If you find it difficult to visualize the colors, simply see them as brilliant white light. In turn, you can visualize the negative energy as something black in color. It is actually easy to memorize the arrangement of the colors of the chakra from the root chakra and up to the crown chakra. If you can notice, the 7 main chakras actually follow the 7 colors of the rainbow (ROYGBIV). Red is for the root chakra; orange is for the sacral or sex chakra; yellow is for the solar plexus chakra; green is for the heart chakra; blue is for the throat chakra; indigo is for the third eye chakra; violet is for your crown chakra.

Chapter 9: Psychic Abilities

What Are Psychic Abilities?

Psychic abilities are another interesting topic in spirituality. It should be noted that when you practice the exercises in this book, especially the meditation techniques, then you will definitely awaken your psychic abilities. This is because there is really nothing supernatural about psychic abilities, especially when your chakras are strong and healthy. There are many kinds of psychic abilities. Let us discuss them one by one:

- Clairvoyance

Among the different psychic abilities out there, the psychic ability of clairvoyance is probably the most common. Once again, clairvoyance is the ability of clear-seeing. We have already discussed this when we talked about the intuition as the power of clairvoyance relates to the third eye chakra. But, just to give you another effective exercise that you can do, here is something that you can try:

Assume a meditative posture and relax. Think of a place in your house. Make it specific, for example, your bathroom or the other room in your house. Now, visualize it as clearly as you can. Pay attention to small details such as the arrangement of objects, the floor, ceiling, walls, and others. Take note of what you see. Now, open your eyes, and check if you were able to "see" the said room in your mind's eye.

This is an excellent exercise that will allow you to travel and see places only with the power of clairvoyance. For this exercise, you can also use the visualization screen that we have talked about. Do not rush in doing this technique. Take as much time as you need to take note of the details in your mind's eye.

Now, in your first several attempts, you will most likely fail to see everything clearly and correctly, but do not allow this to discourage you. Instead, just maintain a positive mindset and keep on practicing. Do not allow failures to discourage you. You will never truly fail as long as you keep on trying. The more that you practice this technique, the more that you will get good at it. This is simply how it is when you try to learn a psychic ability. It is just like learning any other new skill. So, just keep on practicing and doing your best. After some time, as long as you are patient enough, you will soon notice some progress. You will start to see that you are more able to envision the place in your mind more clearly and that you can tell correctly where certain things are as if you were actually in the place that is being visualized. All these use the power of the third eye. If you train yourself well, then this will allow you to see different places and dimensions. Although this may seem like a simple visualization exercise, it is also very effective. In fact, if you come to think about it. All of these spiritual practices are actually very simple, but it just a matter of practicing it sincerely and continuously that can create a big difference.

- Pyrokinesis

Pyrokinesis is the psychic ability to control fire. Yes, there are some people who can do this. Unfortunately, there are many scammers and hacks out there that use trickery to do a similar act. The truth is that pyrokinesis is real, and you can control fire with your mind. Here is an exercise that will allow you to do this. For this exercise, you will need to use some matches:

Assume a meditative posture and relax. Now, light a match and focus on the flame. You should try to connect to this flame and make it "jump" out of the matchstick. There is no right or wrong way to do this. You have to find your own way. The key is simply to focus on the flame and then be one with it. Once you are with it, you will feel as if the flame has become an extension of yourself. This is the time when you can actually control the flame as if it were your arm or finger as if it were just another part of your body. You can now make it jump or move it to be separated from the matchstick.

This technique may take some practice. If you are just starting out, you might find it a challenge. In fact, out of 20 tried, you might only be able to do it once or even none at all. However, do not be discouraged and just keep on practicing. The more that you practice, the more that you will get good at it. This is just the first basic exercise for learning pyrokinesis. Once you get good at it, you can now try another exercise where you get to control the flame and make it dance; hence it is called as the dancing flame exercise. For this exercise, you will need a candle:

Assume a meditative posture and relax. Light a candle in front of you and just focus on the flame. Be one with the candle flame. The more that you focus on the candle, the more that you will feel as if you were being one with it. Try to associate yourself with the candle or at least make it as an extension of yourself. The key here is to focus on the candle flame and not to allow other thoughts to exist in your mind. Now, once you feel like you are attuned with the flame, will it to bend to the right (or to any direction of your choice), and then make it bend toward another direction, and so on. You might want chant, "Bend to the right (or any direction of your choice)," to impress your will upon it. Another technique is to visualize the candle flame bending or leaning in the direction where you want it to bend. Now, if you get good at this, then you can take it a step further. This time, what you need to do is to make the flame move so that it separates from the wick. This is the same as the first exercise with the match. This is the way to extinguish the flame. Another technique is to visualize the flame getting smaller and smaller until it finally disappears.

Now, the next part is probably the hardest part. This technique is a way to light up a candle. The steps are as follows:

Visualize the molecules of the candle. See the molecules moving faster and faster as they heat up. As you do this, see and feel the wick of the candle getting red and hotter and hotter. Continue with this visualization. Believe in it and in its power. Soon enough, with enough willpower, the candle will light up.

This is an advanced technique so do not expect to be able to do it easily. You might want to practice these techniques as a routine. First, try to just move or make the candle flame bend, and then have it separated from the wick to snuff it out, and then finally relight it back. Once you can do all these, you can easily use the fire element in anything. This is how spiritual masters can set an object in fire from a distance. Indeed, this is also not easy to master. But, if you feel like you are attuned to the element of fire, then this is probably the psychic ability that is best for you.

- Telekinesis

Telekinesis, or simply TK, is another famous psychic ability. It refers to the ability to move, or influence injects with the mind. Now, it should be noted that telekinesis is divided into two kinds: micro telekinesis and macro telekinesis. Micro TK refers to the ability to influence randomness or odds. For example, being able to influence a random number generator or a shuffled deck of cards. Macro TK refers to how most people understand what TK is, which is the moving of objects with the mind. For you to better understand micro and macro TK, it is best to do so using an actual practice:

Micro TK

For this exercise, all you need is a coin. Remember that micro TK is about influencing the outcome of randomness or odds. Here, you will use your mind to influence the outcome of a coin flip. The first step is to get a coin – any coin will do as long as it is well balanced, which means that when you flip it, it will randomly be a head or a tail side. The weight should be distributed equally on both sides. Do not worry; it seems that all ordinary coins are like this, so just use your regular coin. The second step is to choose your side: head or tail side. What you will do is to influence your chosen side so that it will come up with every toss or flip of the coin. Finally, you have to influence the said coin as you flip it 100 times.

Now, in theory, when you flip a coin 100 times, then the final outcome should be 50 heads and 50 tails, or somewhere close to it. But, if you can apply micro TK, then it should show a significant difference, like 70 head side and 30 tail side, if head is your chosen side.

So, how exactly do you do this? How should you apply micro TK? Well, the best way to do it is with your mind, and that is by using your will and visualization. The steps are as follows:

Relax. Telekinesis is more effective when you are relaxed. Realize that you are not forcing something to happen. Rather, you are making or letting it happen. Now, let us assume that your chosen side is the head side. As you flip or toss the coin, visualize in your mind that it falls with head side up. As you do this, exercise a strong will, as if commanding the coin to do as you will.

Here is another technique that you can use:

Hold the coin in your hand. Now, just stare at it and focus on it. Be one with it as you try to be one with the flame. Just focus on it until it feels as if it has become a part of you. Now, once you feel that you are in control of it. Just use your willpower to influence it and make it give you the outcome that you want. As you flip the coin, feel that the coin is a part of you and just make it land the way you want it.

There is really no secret to this technique. It is all about becoming one with the coin and controlling it to make it do your will. Do not identify the coin as a separate object. Rather, you ought to feel as if you were the coin or that it is at least a part of you, like your arm or elbow. Be one with the coin. You are the coin.

This is just an example of how micro TK can be used. If you reach a point where it becomes easy for you to control the outcome of a coin flip, you can try other things, such as a shuffled deck of cards or a dice. If you are using a shuffled deck of cards, an exercise that you can try is to force a color. For example, think of a color (red or black) as you shuffle the deck of cards. The top card of the shuffled deck of cards should be the chosen color. Feel free to device your own way to practice micro TK. Once you acquire and develop this ability, you can use it for many purposes. In fact, there are those who claim that this can be used to influence the odds in the casino or even the lottery. You might be surprised, but there are real-life lottery winners who are psychics and spiritual practitioners. Take note, however, that psychic abilities are not really meant to make you rich. Instead, you should use them to help other people and grow more spiritually. Never lose the real purpose for practicing these things, and always cling to what is good.

Macro TK

Macro TK is exactly what most people understand when they encounter the word, telekinesis, and this is actually moving objects with your mind, such as moving a coin, ring, or even heavy objects like a television or a car. You might be wondering, is this really possible? Well, this might surprise you, but the answer is a resounding *yes*. You see, there is no limit to the power of your mind except the one that you have made against it. The following exercise will help you practice and experience macro TK:

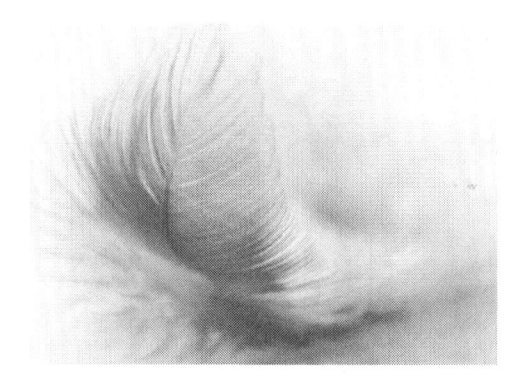

Place a light object in a table in front of you. Now, focus on this object. Be one with it. Focus on it to such an extent that you see nothing but the object. Create a tunnel in your mind that connects you to the object and let nothing else exist but this. Feel and be one with the object. Now, in your mind's eye, see your astral hand reaching out and pushing the said object.

Did it move? Another way to do this is to simply use your willpower and just be one with the object as you make it move with your will. Now, most people may require the use of visualization. You can try both ways and see which one works best for you. Again, do not be discouraged if nothing happens on your first few attempts. Just keep on practicing.

Here is another exercise that you might want to try. This can help you to levitate something, which is also a part of macro TK. The steps are as follows:

Place a light object, preferably a feather, on your hands. Keep your hands in front of you with the palms facing upward. Feel the object resting on your palms. Feel how light it is. Now, we will use energy to levitate this object, and we will draw energy by using this chant: light as air, light as a feather. Keep chanting it for a few minutes as you connect to the object and will it levitate.

Instead of using a feather, you can also use aluminum foil. The aluminum foil might not levitate, but it will most probably move on your hand.

Take note that telekinesis, especially macro TK is considered an advanced psychic ability, so do not expect to be able to do these exercises immediately without spending enough practice. Still, this ability is something that is worth learning. Once again, do not allow it to make you lose your sight on what is truly important in your spiritual life. Remember that although acquiring psychic powers can be nice and interesting, it is not the end of spirituality.

- Hydrokinesis

Hydrokinesis refers to the ability to control water. You might see some videos on this on YouTube. Unfortunately, many of those people who promote themselves on such sites are only looking for attention and their abilities seem sketchy or fake. So, the best way is still for you to try it out yourself. For this exercise, you will need a basin and a needle. The steps are as follows:

KUNDALINI AWAKENING

Fill the basin with water. Now, place the needle in the basin and let it float. Take note that hydrokinesis deals with water so you would want to focus on the water. The needle is only there to show you if what you are doing is working. Now, visualize pushing the water with your energy. You might want to mimic the action with your hand as if you were pushing it with your hand, although you are actually doing so only with the power of your mind. Push the water to one side. The needle should respond and be pushed towards that side/corner. Focus on the water. Be one with it and push it. You can also visualize a wave in the water and use that wave to move the water. Just realized all throughout and connect with the element of water.

Once you get good at this, you can take it a step further. Go to the sea or any body of water. Now, use your ability to manipulate the water. It is believed that once you master this technique, you can do wonders such as changing the course of a river or of the typhoon. You can actually be able to control water and direct its flow. However, just like any other interesting psychic ability, this also requires practice.

- Clairsentience

Clairsentience, also known as clear-eyed information, is the ability to feel the subtle energy. If clairvoyance is more about the sense of sight, then clairsentience deals with feelings. This is not as hard as you might think. In fact, you probably have a good sense of clairsentience, and you are simply not aware of it. A common example of this is having a bad feeling and then something undesirable happens, such as an accident. Or, you might feel a positive energy only to realize that an angel just passed by. This is more about feeling, not just energy but also the quality of energy. So, how do you develop or practice this ability? Well, just keep on feeling energy. The best way to do this is to develop your empathic ability. You can use this exercise:

Make sure that you are in a public place, or anywhere where there are people around you. Now, look around you and try to find someone whom you feel is a nice person. You do not have to be logical about this, but let your intuition decide for you. Now, just focus discreetly on this person. Feel your heart chakra. Remember that your heart chakra is the center of emotions and universal love and oneness. Now, see and feel a ray of light shoot off from your heart chakra and let it connect with the heart chakra of the chosen person. This will serve as your link and connection to the person. The next step is to simply clear your mind. You should be able to receive impressions or thought and ideas, even emotions, from the person.

You are not limited to feeling only the energy of people, but you can also use your ability to sense the energy of a place. If you do not like to travel, you can even conjure the place in your mind's eye and try to feel it from there. Once again, just like the other skills, this one also takes practice. The key here is to get used to using your feelings to sense the subtle energy around you. The more that you practice this, the more sensitive you will be to the subtle energy.

- Divination

The practice of divination is very common to those who are into spirituality. A common tool that is used by dowsers is the tarot cards. Now, this book will not define every tarot card as that can be a lot. Instead, you will learn how to use any kind of tarot decks. The way to do this is by using your intuition. So, how do you do this? Well, the truth is that all tarot cards communicate a message that you can decipher. Instead of giving each card a meaning that the manufacturer has provided, you can also come up with your own personal meaning. The way to do this is by using your intuition.

KUNDALINI AWAKENING

Now, there are no hard and fast rules on how to do this. You will want to pick a card or make a spread. Once you have the cards that you need to interpret, just relax and clear your mind and allow your intuition to take full reign. Look at the card and focus on it. How does it make you feel? Do you sense any ideas or emotions being evoked by the card? Take as much time as you can to understand the card. With enough practice, you will get good at this and be able to read any tarot cards. You should realize that tarot cards are just tools. They do not form rules or regulations. Instead, you use them to communicate to you a message. It would be impractical to memorize all the meanings of all the tarot cards, especially now that there are countless tarot cards out there.

Another method of divination is by using a pendulum. In this case, you will have to learn the art known as dowsing. Okay, what is a pendulum? A pendulum refers to any object that is suspended on a string or chain. Before we proceed with the instructions, you first need to acquire your dowsing tool, a pendulum. Now, many occult shops sell a pendulum. However, if you are just starting out, you can simply create your own pendulum. All you need is a string or thread, and a needle, the ones used for sewing clothes would be enough. Simply cut the thread to your desired length, about a meter would be enough for a pendulum. Next, tie one end to a needle. You can now use it as a pendulum.

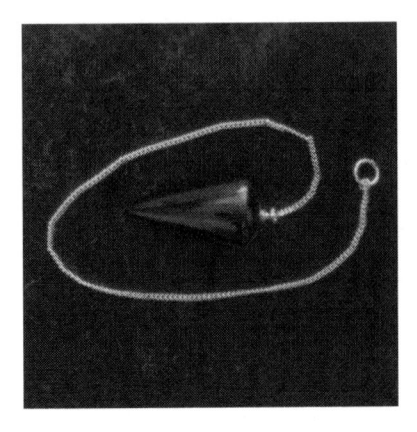

When used in dowsing, a pendulum can give you answers to your questions, specifically questions that are answerable by *yes* or *no*. For example, you can ask and dowsers, "Is it going to rain today?" As you can see, a pendulum can be a very effective and versatile tool that you can use. Now that you have a good idea of what dowsing with a pendulum is about, let us move on to the instructions:

The first thing that you want to know is how your pendulum responds for a yes answer and for a no answer. The way to do this is simple: Simply hold your pendulum by the end of the strong and allow it to hang freely. Once it is still, say, "Show me yes." The pendulum should move to signify a yes answer. Take note of it. Now, command it to, "Show me no." and pay attention to how it moves for a no. Do not stop until you can take note of the appropriate response that you are looking for. Needless to say, the pendulum should give you two different movements – one for no and another one for a yes answer. Once you know how your pendulum responds, you can now start asking it questions. Do not rush with hard questions right away. You have to warm up and get used to it. Hence, it is advised that you first ask your pendulum questions to which you already know the answer to, such as, "Am I wearing a jacket?" or "Is it Sunday today?" Check if the pendulum gives the correct response. Once you get used to this, then it is time to use your pendulum to answer questions the answers to which you do not know of. Here are the instructions:

Allow the pendulum to hang freely as you hold it by the end of the string. Let the bob of the pendulum (the needle) hang freely and let it become still. Once it is still, you can now ask your question. Be sure to frame your question correctly. This means that it has to be answerable by yes or no. You should also keep it short and clear. As you ask your question, focus on it. If the pendulum does not give you an answer, simply repeat your question until the bob moves. Feel free to ask your pendulum as many questions as you want, keeping in mind the proper frame of questioning.

Dowsing is a skill that takes practice. So, do not expect to get all the right answers in the beginning. You simply have to practice more. It is also advised that you use only a single pendulum and to make a good connection with your pendulum. To cleanse your pendulum of negative energies, you can wash it in running water or with saltwater. You can also let the rain wash it for you. Dowsing is a very interesting practice. It is often used to find water, lost objects, hidden treasures, and even missing people, as well as to get answers from life's biggest questions.

- Psi healing

Psi healing or the ability to heal using prana is another very famous and helpful ability that you can learn. As the name suggests, this is where you heal ailments and illnesses by using prana. Now, there are many ways to do this. The most common schools for this are reiki and pranic healing. Both schools of healing use prana to effect healing but there are just some slight differences. Reiki makes use of secret symbols while prana healing does not use such symbols but focused on no-touch energy healing. It would take another book to discuss reiki and pranic healing in full, but to help give you an idea of what it is like, here is an effective technique that you can use to effect healing. The steps are as follows:

If there is a part of your body that is not well, you can charge it with prana to help facilitate the healing. The way to do this is to simply charge the affected part/s with energy. As you inhale, see and feel pranic energy being absorbed by the affected part. Continue to do this until it becomes filled with lots of prana. Make sure that you have the faith and intention to heal using prana.

It should be noted that the physical body is hard-wired to heal. The problem is that there are many practices in modern time that blocks or prevents the body from fully exercising its latent healing powers. By charging the affected part with prana, you get to help it heal. Take note that healing requires energy. You might want to try this technique when you have a toothache, headache, tummy ache, or others. If you have a fever, you can charge your whole body with prana. If you get good at this, you can even use this ability to heal other people. Indeed, energy healing is one of the abilities that many people want to learn these days. Take note that just like other wonderful abilities, this is also in your power to learn, but it does demand lots of practice, especially the practice of regular meditation.

Now, there are so many other psychic abilities out there. It is simply too many to mention. However, all of these abilities always make use of energy. This makes them all one and the same. It is just how this energy is manifested that makes a difference. Again, it is worth repeating that acquiring psychic abilities is a normal part of the journey the spiritual path but keep in mind that it is not the end that you seek. In fact, in some traditions, they intentionally renounce psychic abilities thinking that they would only hamper or be an obstacle to true initiation and enlightenment. Of course, this is a matter of personal belief and preference. It still depends on how you manage and use those abilities. This book suggests that you should not ignore your psychic gifts. Instead, you should learn to use them properly to help people and always for what is good.

Psychic Awareness

Psychic awareness happens when you develop your chakras. Your chakras function as the senses of your spiritual body just as the physical body has its different parts for certain functions, so do you use your chakras in the spiritual or astral realm. Of course, when it comes to being aware psychically, that has more to do with being aware of prana or energy. Again, everything is made of energy. By doing the meditation practices in this book, you will surely be able to be more sensitive to this energy, which will increase and enhance your psychic awareness.

When it comes to being aware, you would want to work on your intuition. Hence, you have to develop your third eye. Just as it is easier to be aware of your physical place by using your physical eyes with its sense of sight, you can make use of your third eye chakra to be more aware of the energy around you. Of course, there are also other senses that you can use.

Now, being aware is one thing, knowing what to do is another. You have to be careful about this. Many practitioners learn to be aware but then do nothing. When you do nothing, then it is as good as if you were not aware of anything. The problem why this mistake happens is that many people still fail to trust in their intuition. For example, when you feel a sense of danger about a particular place, you should leave that place immediately without having to wait for something bad to happen. Learn to trust your intuition. Not trusting in your psychic awareness is one of the effective ways to start losing it. It is believed that all human beings used to have a strong psychic awareness. However, most parents would train their children to focus more on physical objects, and so people tend to forget about their natural ability to be psychically aware when they grow up. Well, this should not be the case. Now is the time for you to remember other senses (psychic senses) that you have and start learning and using them once again.

Enhance Your Psychic Abilities

Okay, so how do you enhance your psychic abilities? Well, this is simple: just practice. The more that you practice, the better you will get. This is also why many people fail to achieve success. Practicing for a long time is not easy. Most people would give up halfway and just abandon their spiritual practice. You see, when it comes to learning a new skill, doing lots of practice is often required. For example, let us say that out want to learn how to juggle four balls at the same time. Even if you are given a clear set of instructions on how to do it, you will most likely fail to do it successfully on your first few tries, and perhaps not even after a week. Even if you can memorize the instructions word for word, there is no guarantee that you will be able to execute it well. The same applies when you learn anything psychic ability. Although knowing the instructions is important, it is not enough. Another very important element is to practice continuously. As you can see, there are two elements to your success: the right knowledge and continuous practice. This book provides you with the right knowledge; it is up to you to turn that knowledge into actual practice.

The practice of meditation cannot be overrated. When it comes to developing your psychic abilities, regular practice of meditation, even of the basic breathing meditation, can be a very big help. You have to embrace the practice of meditation. Make it a priority to meditate every day.

Chapter 10: Astral Travel

Understanding Astral Travel

Though astral travel is still a mystery to some people, its practice has become very famous. Although there are those who still see it as something strange, you should know that you, as well as all other people, always astral travel when you sleep. The problem is that only a few can remember what happens on their astral journey. Okay, just a quick review: what is astral travel? Astral travel is where your astral body (people refer to this as your soul or your consciousness) separates from your physical body and travels the world. This is, indeed, an interesting ability and one that will allow you to travel the entire world for free. Yes, this will allow you to go to the pyramids, see the Bermuda triangle, go to Italy, Japan, and anywhere you want – even the moon and other celestial spheres as well as magical planes of existence. This is why so many people want to learn how to astral travel.

When people talk about astral travel, a very common question that you will hear about is this: Is it safe? Okay, this is a legitimate question. Since the astral body separates from the physical body and death is nothing more but the permanent separation of the astral body from your physical body, then it appears that astral travel is dangerous, right? I mean, it can kill you, correct? Well, not really. In fact, astral travel is safe, especially if you know how to do it properly. Do not be afraid of it. You do not have to. Just think about the fact that you have been astral traveling for years already, but are simply not aware of it. Learning how to astral travel only allows you to be aware and be in control of your astral journeys.

Fear is a very common subject when learning how to astral travel. Even though you are told that it is safe, it is not easy to just believe it and not be afraid. How can you not be afraid when you see that you are actually separated from your physical body? Well, it just takes practice until you get used to it. In fact, it was found that fear is one of the reasons why people fail to astral travel. This is because when you experience fear, you assume a defensive position. The tendency here is that your astral body will be pulled back into your physical body, which prevents astral travel or prevents it altogether. Hence, you must learn to conquer that fear. You have to realize the truth that astral travel is safe. Once you feel afraid, even if it is a groundless fear, you will be pulled back immediately to your physical body. Also, going back to your physical body is very easy to do intentionally. You simply have to think and will yourself back to your body. Many times, just the thought, "I want to go back to my physical body." would be enough to pull you back, so you need to be careful with your thoughts. The key is not to be afraid and adopt a positive mindset. Moreover, having a negative mindset while you are in the astral plane is not good since you will probably attach negative entities. Again, this is not something for you to be afraid of as you will easily be pulled back into your physical body before any real danger can get at you.

When you astral travel, you might also encounter the silver cord. This silver cord is like the umbilical cord that is often attached to a new-born baby. Well, you might find one while in astral in the form of a silver cord that attaches your astral body to your physical body. This only goes to show that your astral body is not completely separated from your physical body even while astral traveling. Take note that you may or may not see this silver cord. If you do not see it, then that is not a problem. But, if you do see it while you are astral traveling, then now you know that it is normal. Just ignore it and focus on your astral journey.

When you astral travel, you might be able to see yourself and even see your astral body, but you may also appear to be nothing but space or consciousness. You may be able to see around you, but it is also possible that you can only hear things. Do not worry; the more that you learn how to astral travel effectively and the more practice you get, the better you will be.

It is also possible that you meet other beings or entities in the astral plane. However, if you are just starting out, it is advised that you do not commune with anything or anyone in the astral. Do not worry, as long as you do not pay attention to them, then they will not bother you are well. Again, do not be afraid. If you ever feel like danger might be near, you can always go back to your physical body; and even in the case where there is a real danger, you will automatically be pulled back to your body as if awaking from a dream.

Not only will the practice of astral travel allow you to visit wonderful places, but it can also make you realize that you should not fear death. Death is nothing but the mere separation of the astral body from the physical body but in a permanent manner. When you astral travel, this only happens temporarily. Still, it is a relief to know that there is indeed life after the physical body perishes.

Today, many people want to learn how to astral travel due to its wonderful benefits. The good news is that anyone can learn how to astral travel. Yes, you can learn how to do it, too. But, just like any other skill or ability, being able to astral travel properly requires some practice. Luckily, this is easy to do as you can always try to do it when you go to sleep. When you astral travel, the body goes to sleep, but your consciousness remains awake. Now, there are many ways to do this. Let us examine notable methods of astral traveling.

How to Astral Travel

- Mental travel

Okay, this is not really astral traveling, but it is a nice first step to learning how to astral travel. In fact, once you get really good at this, it can lead you to a full-blown astral travel. As the name implies, this one only uses your mind and not your astral body to travel. But, if you have very powerful visualization skills, then this can be an astral travel the more that you go deep into it. The steps are as follows:

Lie down and relax. Breathe in and out and just relax completely as if you are going to sleep, but be sure to remind mentally awake and to keep your spine straight. Now, visualize the room that you are in. see and feel as if you are looking through your closed eyelids with your mental eyes. Now, just try to look all about the room and pay attention to the objects and small details. Now, visualize the next room, then the next until you reach outside of your house. Now, feel free to roam and see the wonderful neighborhood and go anywhere you like. When you are ready to end this exercise, simply visualize the room where your body is in and then think of your physical body. Slowly move your fingers and toes and open your eyes with a smile.

Indeed, this is nothing but a mere visualization exercise, but it is also very effective. It is important that you try to visualize the places as clearly as possible. Again, this is not astral travel as you do not need to visualize places when you are in a real astral journey, but this is nonetheless a good preparatory step to learn and experience astral projection.

- Roll out method

This is a popular way to do astral travel. Here, you separate your astral body from your physical body by rolling to one side. The steps are as follows:

Lie down and relax as if you were going to sleep. Indeed, your body will fall asleep, but your consciousness will remain awake. Now, just relax and allow your tired body to rest and sleep. Think and feel that you are an astral body that is locked in a physical body. Look around you with your astral eyes while your physical eyes are closed. Now, just maintain this mindset for several minutes. Once you feel more connected to your astral body, believing strongly that you are the astral body and not the physical, you should be able to feel your physical body to be asleep, but you should remain conscious in your astral body. Once you reach this stage, you can now roll out to the side with your astral body. If you can do this well, then you will be able to roll to the side and be present in your astral body. Do not be afraid to see your physical body sleeping right in front of you for even a slight fear can end an astral travel and pull you back into your physical body. Remember to only roll when you feel that you are np-longer associated with your physical body but becomes more attached to your astral body. You might have to try this several times to know the right feeling or signal that it is time for you to do the rolling motion. Feel free to roll to either side that you want, the important thing is to do it with your astral body.

- Floating method

With this technique, the idea is to float with your astral body. The steps are as follows:

Lie down and relax. Be sure that your spine is straight. Just breathe in and out and relax completely but do not fall asleep, only your body should be put to sleep. Once again, see and feel not as the physical body but as the astral body trapped in your physical body. Look around you with your astral eyes. You might want to use your visualization skills in the beginning as you try to be more connected to your astral body. Allow your physical body to rest and fall asleep. Now, see and feel that as you inhale, your astral body gets lighter and lighter. As you become lighter, you start to float. First, you float above your body, and then higher and higher. Stop thinking about your breath. Instead, put all of your focus on your astral body. Fully identify yourself as the astral body. Now, continue to rise higher and higher. Rise above your house and up the clouds. What do you see? Now, you are now in the astral. In the astral, all you need to do is to think of a place, and you will be transported to that place in an instant. So, where do you want to go? Just think about it, and you will be there. If you cannot think of a place, simply roam around your neighborhood or any other place while you are in your astral body. When you are ready to return to your physical body, first think about your room. Once you reach your room, watch how your physical body sleeps sweetly. Walk close to your physical body and enter it. Slowly move your fingers and toes and gently open your eyes.

- Through the third eye

This technique will allow you to separate your astral body from your physical body through your third eye chakra. The steps are as follows:

Lie down and relax completely. Allow the body to fall asleep but keep your mind awake. Make sure to keep your spine straight to facilitate the flow of energy through the chakras. Now, once you are feeling very relaxed, identify yourself as the astral body and not the physical body. See and feel how your astral body is trapped in the physical. Now, visualize your third eye chakra glowing and consider it as some type of a gateway. This gateway will lead you into the astral realm. Continue to focus on your third eye chakra and continue to think of it as a gateway to the astral. Now, move your astral body and escape through your third eye chakra. Once again, do not be surprised to see your physical body lying in front of you. Remember that the third eye chakra is the chakra that is closest to the spirit or astral world, so it is an excellent place to enter the astral realm.

You can try all of these techniques for astral travel and choose the one that you prefer the most. You do not need to learn all of the techniques as you can use the same method again and again to astral travel. The important thing is to use a technique that works best for you. You can also make adjustments to these techniques as you may seem fit. After all, there are no hard and fast rules as to how you can astral travel. In, fact, most astral travels happen unintentionally where you just get conscious halfway on the journey. However, these techniques are used to deliberately and intentionally cause an astral travel.

Tips

Let us now discuss notable tips to help you astral travel successfully:

- Relax

It is impossible to astral travel if you do not relax. If you do not relax, the more that you will be attached to your physical body. Hence, be as relaxed as possible. In fact, when you astral travel, it is good to have your physical body to be asleep with your mind active and conscious. In the initial stage, do not even think about astral traveling. Instead, you can focus on relaxing your body as if you are going to sleep, just be sure not to let go of your mind that you might lose consciousness. Relax and let go.

- Do not expect for anything

Having expectations can ruin an astral travel. You should let go of any expectations. Do not even expect to be able to do it. Expecting will only have your energy to be divided. Not to mention, it can add more pressure, which is counterproductive to your efforts. Instead, of having expectations, you should relax more and just be conscious. Although you may know what to expect when you astral travel, it is good to forget about all you know when you actually engage in the induction of an out of body experience. Just focus on what needs to be done and forget about everything else that you know.

- Regular practice

Of course, an important part to success is doing regular practice. You can practice astral travel before you sleep at night. However, if you face issues with always falling asleep and losing consciousness, then it is advised that you try to astral travel early in the morning. This way you will have already rested, and there will be fewer chances of falling asleep.

- Empty stomach

It is recommended that you try to astral travel on an empty stomach. Astral traveling when you are full causes your energy to be divided since your physical body will use energy to digest the food, so avoid astral traveling after you have just eaten. Give it some time off at least two hours, especially when you have eaten a heavy meal. Light snacks are okay, especially when you are feeling hungry just to satisfy your stomach, but do not eat a large meal. You want to conserve your energy and use it solely for astral traveling.

- Psychic protection

Although not really a requirement, it is also good to use some kind of psychic protection, especially if you still think that astral traveling is risky or dangerous. You can use the following protective measures:

- Place a glass of salt water beside your body or on the bedside table. It is believed that saltwater does not only cleanse but also repels negative energies.
- Say a prayer before you astral travel and ask for protection.
- Visualize a ball of white light surrounding your body. Know that it is protecting you from all negative energies. It is a shield that protects you from all harm.
- Draw a circle around your bed with salt.
- Sprinkle saltwater on your body before you try to astral travel and know that it shall keep all negative energies away.
- Never entertain negative thoughts while you are astral traveling.

Chapter 11: Connect to Your Higher Self

Understanding the Higher-self

In some traditions, the human being is divided into three selves: the younger self, the middle self, and the higher-self. The younger-self refers to the subconscious mind. The middle-self refers to ordinary consciousness, while the higher-self refers to the god-self. As you can see, it often gets tricky when it comes to the higher-self. Other traditions believe that the higher-self is the best version of who you can be, while others claimed that it is you when you act as a good as you truly are.

Regardless of what you believe in, the truth remains that you have a god-self. The question now is how can you connect to your god-self and how do you benefit from it. Well, the answer is actually simpler than you might think, and that is through meditation. When you meditate, you get to still the mind. Once this internal chatter stops, then it paves a way to the realization and connection to the god-self. This is how gurus and monks get answers simply by sitting in meditation without thinking of anything. In the end, you will realize that you just know. This is your higher-self at work, that part of you that is always connected with the Divine and knows everything.

Unlike the younger self that wishes to be engaged with elaborate rituals and stuff, the higher-self seeks silence and serenity. The more that you still your mind, the more that the higher-self can manifest. Hence, the practice of meditation is a must. You do not even have to desire anything. Simply let things unfold as they are. As the saying goes, "Just be."

Gain Wisdom and Clarity from Your Divine Self

The practice of meditation naturally allows you to gain wisdom and clarity from your divine self. In fact, any meditation practice can do this as long as you do meditation regularly. However, it should be noted that there are also techniques that primarily focus on this purpose. The following meditation practice is a good way to actively gain wisdom and mental clarity by communing directly with your divine self or higher-self:

Assume a meditative posture and relax. Once you are feeling relaxed, try to envision what you think as the best version of yourself. See it not only on the physical level, but on all levels, including mental, emotional, and spiritual. Visualize this version of yourself standing right in front of you and is ready to have a conversation with you. Take note that this being in front of you now is a highly evolved being and can actively communicate using telepathy, so pay attention to thoughts that arise in your mind. Do not force thoughts to appear in your mind; just let them pass freely without any pressure. Now, focus on your higher-self. Start to communicate with it in your mind. Talk to it telepathically. Be open to what it wants to tell you. Enjoy the moment and learn.

Now, realize that this higher-self that you adore so much for its brilliance and wisdom is actually a part of you. Yes, you are this fantastic. You simply have to recognize your own beauty. Now, as you inhale, see and feel that you are inhaling this higher-self back into you. As you do, feel that everything you that admire about this higher-self is now becoming a part of your persona. You are the higher-self. Breathe in and out. Appreciate your own power. You are wonderful.

Feel free to make some adjustments to this meditation as you might prefer. This is an excellent technique that will allow you to talk with your higher-self and also to realize your own power.

Chapter 12: 5-Minute Meditation Sessions

The following meditation can be applied for about 5 minutes, but you are also free to use them for as long as you would like. These are excellent meditation techniques for beginners, and they can have profound effects. The more you practice these meditation techniques, the better you will get.

- Affirmation meditation

You are probably aware of the use of affirmations. For example, when people are feeling afraid, you might hear them say, "I can do this." It is like affirming what they want to happen. Although this is something that is very common, the truth is that only a few know how to use it properly. What you need to know is that affirmations are more effective when they are recited in a meditative state. But, before we discuss them meditation technique itself, you should know how to create your own affirmation. Here are the steps:

Keep it short and clear

You should keep your affirmation short and to the point. As a rule, try to make it just a short and single sentence only. Take note that you will be reciting, almost like chanting, your affirmation, so do not make it too long. Just around less than 10 words would be nice. Also, avoid using hard to understand words. Instead, use simple words that are easy to understand. Examples: I am strong, I am courageous, I am feeling better every day, I am happy, I am healthy, I am getting stronger, and the likes.

Use the present tense

When you make an affirmation, you should use the present tense. Do not say, "I will become a clairvoyant." Instead, you should say, "I am a clairvoyant." Consider this as some kind of trick of the mind, if you would. The reason here is that if you use the future tense, then it might happen only after so many years; and if you use the past tense, then it means that it no longer needs to happen. Therefore, you should use the present tense, to make it manifest right now or at least as soon as possible.

Believe

It is also important that you believe in what you affirm. Without faith, then it would not be of any good. But, as the saying goes, "With faith, nothing is impossible." Therefore, believe in what you affirm. Believe that it has been realized already, and it shall come true. Again, consider this some form of a trick of the mind, but this is how the universal law works and is the secret to make your desires turn into reality. If you do not believe in what you say, then the affirmation loses its power.

Repeat

While you are in a meditative state, you should repetition your affirmation as many times as you may need to make it sink into your subconscious. Use it as a kind of mantra or focus of your meditation. Let your mind absorb it and sink into its meaning. Let your affirmation be the sound of the universe at that moment.

Only use one affirmation

It is not good to use different kinds of affirmations at once. Just focus on one affirmation, and do not change it until you have achieved its objective or if you are ready to just give it up. Using more than one affirmation at the same time can be confusing, and your mind might now know which affirmation to absorb fully. Hence, do it one at a time.

Now that you know the important points about making an affirmation, it is time to move on to the actual meditation process:

Assume a meditative posture and relax. Be sure that you already have an affirmation that you want to use. Do any basic meditation, such as the breathing meditation. The objective is simply to reach a meditative state. Once you reach a trance or meditative state, start saying your affirmation. Use your affirmation as the point of focus of your meditation. Focus on it and be one with it.

When you are ready to end this meditation, simply stop saying your affirmation and just bring your attention back to your physical body. Slowly move your fingers and toes and gently open your eyes.

- Mantra meditation

The mantra meditation is another very popular and powerful meditation technique. What is a mantra? A mantra is a sound, word, or syllable that acts as the point of focus in meditation. It helps to silence the mind as well as to evoke certain toes of energy. For this meditation, we are going to use the mantra, OM.

The mantra OM is very famous and powerful. It has been used by many spiritual masters and monks. It is also very common in

Buddhism and Hinduism. It is believed that OM was also the very first sound in the universe. When you use the mantra OM, you do not just identify yourself with others who meditate, but you also tap the energy of many other spiritual masters and gurus in the world. Let us now move to the actual meditation proper:

Assume a meditative posture and relax. Now, start to say your mantra. In the beginning, you will have to say it out loud. However, after some time, the mantra will be a natural part of you that all you will need to do is close your eyes, and you will be able to hear it with your inner ear (clairaudience or clear hearing). This will happen once you get used to your mantra and have established a good connection to it. Keep your focus on your mantra. Relax and follow your mantra.

As you can see, this is a very simple technique, but it is also very powerful. This is why it has been a long time favorite among meditators, beginners as well as well-experienced ones.

If you are not comfortable with using the mantra OM, you are free to use other mantras. You can even come up with your own mantra. However, when it comes to making your own mantra, you need to take note of some important points:

It has to be neutral

The mantra must not evoke an image or anything. For example, it is not good to use the word elephant as a mantra since it will make you imagine an elephant, which can cause your focus to be divided. Instead, choose a mantra that will not make you visualize anything so you can focus on it without a problem.

Easy to recite

You will have to say your mantra countless times, so be sure to use one that is easy to pronounce. It is also recommended to use a short mantra for convenience.

Use it many times

To establish a good connection with your mantra, you should use it many times. A good advice is to say your mantra even when you are not engaged in an actual meditation. For example, while driving or cooking. The point here is simply to get used to it and make it more a part of you. The more that you become closer to your mantra the more effective it will be. Just like with the use of affirmations, it is advised that you stick to using the same mantra. Hence, early in your spiritual journey, you are not encouraged to make time to try and choose the right mantra for you.

It is worth noting that a mantra does not need to mean anything. Its primary purpose is to help you still the mind by making the mind think of only a single thought (the mantra) instead of having too many thoughts (the monkey mind). Remember that your mantra should help you focus and still your mind.

- White light meditation

This is a good meditation technique especially if you want to trigger a lucid dream, also known as conscious dreaming. It is also a good way to just give you a relaxation as it will tend to make you fall asleep. The way to do this is as follows:

Close your eyes and relax. Consider everything that you see that is not black as light. Now, focus on the light. Be as relaxed as possible, even fall asleep, but keep your focus on the light. Let go.

What will happen here is that you will most likely start to see images after some time. These images will soon turn into a vision almost like a dream, but you will be conscious of it. Hence, you can take control of your dream. This exercise can even lead to an actual astral travel once you get good at it. Needless to say, it is not good to expect to see visions for reasons that we have already discussed. Instead, just focus on the light and let go.

- Energy charge

This is an excellent technique to get a boost of energy when you need it. This meditation technique fills your whole body with prana. Here are the instructions:

Assume a meditative posture and relax. Now, visualize a ray of white light descending from the sky and allow it to enter your crown chakra, and then into your body. Let it fill you with divine energy. See and feel your whole system filled with the strong current of energy from above. When you are satisfied and ready to end this exercise, simply see the ray of white light slowly fade away, and enjoy the boost of strong energy.

If you get good at using this technique, you can use it anywhere, even in public. All you need is to do it with your mind without having to assume a meditative posture. Just draw positive energy from the sky and fill yourself with it. This technique might take practice, but it is nonetheless helpful.

Another way to charge yourself with energy quickly, especially when you are feeling tired is by hugging a tree. As you hug it, feel its energy charging your system. Do not forget to thank the tree afterward.

- Bubble shield

This is a meditation technique that will allow you to create a protective shield around you to protect you from negative energies. If you deal with energies, then this is something that you should learn as you will be more sensitive to subtle energy, including negative energies. The steps are as follows:

Assume a meditative posture and relax. Visualize yourself surrounded by energy. For beginners, you can just see this energy as white light. Now, as you inhale, sea and feel that you draw energy from around you and have it form a bubble shield of protection around your body. With every inhalation, continue to charge your bubble shield. See and feel as it gets stronger and harder with every breath. Affirm, "This bubble shield protects me from all negative energies."

Take note that the strength of this bubble shield will dissipate over time. To keep your bubble shield strong, be sure to absorb more energy to replenish it. On average, a typical bubble shield can last for about five hours. The more that you are exposed to negative energy, the quicker that your bubble shield is going to weaken, so be sure to be sensitive enough about it. If you sense that it is getting weak, then replenish it with more energy. It is good to use this technique before you mingle with people or when you know that you are going to a place where you will be exposed to different kinds of energies, especially if it involves negative energy. It is also noteworthy that this shield gets stronger the more that you get used to it.

- Alternate nostril breathing

This is a breathing technique that is used in yoga. When you use this technique, you will feel a sense of balance and tranquility. The way to do it is as follows:

Assume a meditative posture and relax. Now, press your right nostril on the side with your thumb, and inhale through your left nostril. Hold the breath for about one or two seconds, and then press the left nostril with your ring finger as you release your thumb and exhale through your right nostril. Try to keep a sense of balance by making your inhalation and exhalation of the same length duration. Relax and continue to focus on your breathing.

Although this may seem a very simple exercise, you might get surprised how effective and powerful it is. It will give you a sense of peace and mental balance, as well as clarity. It is not an excellent meditation technique both for beginners, as well as for experienced meditators. Be sure to give this a try.

Chapter 13: 30-Minute Meditation Sessions

By now, you should already have some experience with meditation. It is time for you to engage with longer meditation sessions, which can be more intense than the short 5-minute meditation. Before you practice the following techniques, you might want to read through the instructions first to familiarize yourself with the meditation. Having said that, let us now begin with the meditation proper:

- Full chakra meditation

For this meditation, you will once again meditate using your chakras. Unlike the chakra healing meditation, this technique is not actually for healing, but to learn certain qualities of the chakras that you want. For example, if you want to be able to express yourself better, then when you reach the throat chakra, you will just have to focus on acquiring that specific quality of the chakra concerned. Forget about all its other qualities and just absorb what you need.

Before you start this meditation, it is advised that you have a clear idea of what you want from every chakra. Once this is clear to you, then you can proceed with the actual meditation proper:

Assume a meditative posture and relax. Now, visualize a ray of powerful red light coming from the sky and let it descend and enter your head and down into your root chakra. Allow it to energize the root chakra and absorb the specific quality of the root chakra that you want. Do not stop charging or absorbing energy until you feel that you have fully absorbed the quality that you want. Next, see the ray of light change into an orange color. Now, let it go to the sacral chakra and absorb the specific quality that you want from the sacral or sex chakra. See your chakra being energized with lots of energy. Keep on absorbing the specific attribute or quality that you want. Now, let us move to the solar plexus chakra. See and feel the ray of light turn into a yellow color. Once again, let it fill you and energized your solar plexus as you absorb the quality that you want. Do not stop until you are satisfied. Next, see the ray of powerful light turn into green and allow it to energize your heart chakra as you absorb the specific quality of the heart chakra that you want to learn or develop. Next, do the same with the throat chakra with its blue color. Continue to energize and to absorb. Next is the third eye chakra. See the indigo color shining brightly. Finally, there is the crown chakra. Take as much time quality for the crown chakra as you want. Be sure to really feel it. When you are done, simply visualize the ray of light slowly fade away.

This meditation might seem fast, but it is actually longer than it seems. Take as much time as you need with every chakra. Do not leave it until you have absorbed the quality or attributes that you want from it.

- Earth healing

By healing the Earth, you can also heal yourself. The following technique is based on the teachings of pranic healing. It is about filling yourself with loving kindness energy and sharing it with Mother Earth. The steps are as follows:

Assume a meditative posture and relax. Visualize the earth floating in front of you with the size of a small ball. Now, raise your hands in blessing position, and think of a happy memory. Visualize a ray of white divine light from above descending to your body. Let it enter you from the crown chakra, and let it fill your entire being. This is the energy of pure love and kindness. Feel it like an overflow of positive energy inside you. Now, share this energy of loving kindness with Mother Earth. See and feel white Ray's from your hands and send them to the Earth in front of you. Know that by doing so, you are not just blessing the Earth with positive energy but also all the people on earth. Continue to bless the Earth and everyone. Imagine people turning away from their bad ways and having more faith and love. Imagine yourself feeling so happy and satisfied with your life. As you keep on charging the Earth with positive energy, you are also being filled with loving kindness.

Now, rest your hands on your lap or on the side, and start chanting the mantra OM. Spend about 15 minutes meditating on the mantra OM.

Now, place your hands on the chest level with your palms facing outward. Send the excess energy to Mother Earth.

When you are ready to end this meditation, simply think about your physical body. Move your fingers and toes and slowly open your eyes with a smile.

Another way to do this meditation is to sit on the soil of the Earth. Instead of visualizing the Earth as a small ball, have your hands touch the ground. Just lay your hands flat on the ground and send energy directly to the Earth.

- Forgiveness and blessing meditation

This meditation focuses on giving two things: forgiveness and blessing. Forgiveness is free. You can forgive even those who do not ask forgiveness from you. Forgiveness is a sign of strength and not weakness, for only the strong can give true forgiveness. The steps for this meditation are as follows:

Assume a meditative posture and relax. Raise your hands in the position of giving blessing with the palms facing outwards. Now, visualize a person who has done you wrong. Look at this person. Now, tell them, "I forgive you." Before you let go of this person, see and feel a ray of white light from your hands being sent to the person in front of you. Now, tell them, "I bless you."

Again, think of another person who has done you wrong and repeat the same process. Do not stop until you have exhausted all the people whom you can remember who have done you wrong, especially those who are doing you something bad at this moment or just recently.

When you do this meditation, it is important that you give forgiveness and blessing sincerely. Find it in your heart to forgive. If you happen to find it hard to share forgiveness and blessing, spend some time to think about what is holding you back from sharing positive energy. You should also realize that the act of forgiving is actually good for you. The more that you forgive, the more that you become lighter and even free.

- Meet your spirit guide meditation

This is a meditation technique that will allow you to meet and talk with your spirit guide. It is believed that everyone has a spirit guide. Some people refer to it as a guardian genius or a guardian angel. A spirit guide is one who guides you in your life, especially when it comes to spiritual matters. They will help you in the development of your soul. Unfortunately, many people ignore their spirit guide. With this meditation, you will come face to face with your spirit guide and have a closer relationship with them. The steps are as follows:

Assume a meditative posture and relax. If you want, you can say a little prayer. Also, talk to your guide and tell them that you want to meet them. Now, relax completely. Visualize yourself going down a ladder. The more that you step down the ladder, the deeper you go into trance. One....... two....... three.......... four...... five...... six....... seven....... eight.......... nine.... at the count of ten, you will be in a deep trance....... And, ten....... You find yourself in a beautiful and peaceful place. What does the place look like? Look around you. Suddenly a path appears in front of you. You know that when you walk this path, it will lead you to your spirit guide, the guide who has been looking over you all these years.

Your guide is waiting for you. Walk this path before you. As you walk, you now see a huge clearing in front of you. In the center of this clearing stands your spirit guide. Your guide is happy that you have taken the effort and opportunity to meet them. Come and walk close to your guide and have a chat with them. What does your guide look like? Is your guide even human? Remember that spirit guides usually talk using telepathy, so pay attention to the thoughts that arise in your mind. Enjoy the conversation with your spirit guide.

When you are ready to end the meditation, thank your guide and bid them farewell. Rest assured that your guide will always be with you and that you can always go back to this place to visit your guide at any time. Now, walk back to where you came from down the path and up the ladder. Think of your physical body, slowly move your fingers and toes, and gently open your eyes. Take a moment to think about and reflect upon what happened during the meeting with your guide. So that you will not forget about it, it is recommended that you write it in a notebook.

Chapter 14: Aura

Understanding the Aura

You are probably familiar with the term, aura. But, what exactly is an aura? An aura is an emanation or radiance and plays as a part of the energy body of a person. Now, it should be noted that everything has an aura – people, animals, trees, plants, and even inanimate objects have an aura.

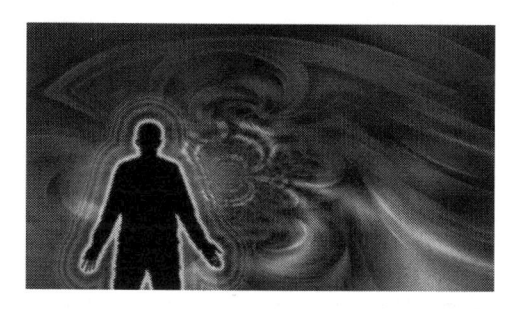

An aura can reveal a person's state of mind, health, as well as emotions. In fact, a certain kind of photography known as Kirlian photography was developed. It was used by a doctor to diagnose their patients. This is because before any illness or disease manifests on the physical body, it will first reveal itself in the aura of the person. The Kirlian photography takes a photo of the aura of a person. But, what you need to know is that there are other ways to know a person's aura. As you can see, it can be very helpful to read auras. To do this, you need to be able to see them or at least feel the aura. Take note that there are different layers of auras. There is what is known as the etheric aura. It is the aura that is closest to the physical body. There is also the mental aura, emotional aura, and the spiritual aura, and others.

How to Feel and See the Aura

Before you learn to see the aura, you must first know how to feel it. Here is a technique which is known as sensitizing the hands as taught by Grandmaster Choa Kok Sui, the founder of Pranic Healing. The steps are as follows:

Rub your hands together. Place your hands in front of you as if you were holding a small ball, palms facing each other. As you breathe in, move your hands slowly apart. As you breathe out, move your hands back together as close as possible, but do not let them touch. Breathe in and out slowly. Soon enough, you will feel some kind of pressure or tingling sensation between your hands. This is a signal that your hands are being sensitized and can now sense the aura. Hold your hands out and try to feel your own aura. You can also walk slowly towards a person or a tree with your palms facing outward. You will soon feel a repelling force or pressure. That is the aura of the person or object, as the case may be. Take note that the more that you sensitize your hands, the easier it will be for you to feel the aura. Try to feel the aura of different people and objects, and you will soon notice their differences.

Okay, but what about seeing the aura? Well, it is not hard to see the aura. For this exercise, you will want to have a neutral background and dim lighting. The steps are as follows:

Hold your hands out in front of you with palms facing outward. Open wide and spread those fingers. Now, set your hands against a neutral background. What you need to do is to apply soft focus. This is nothing more than looking at something without focusing on anything in particular. In effect, you get to use your peripheral vision and capture a wider view. Just relax and stare. Soon enough, you will notice a glow around your hands. It may start as a white light, but the more that you practice you will also get to see other colors like yellow or blue, and others.

Here is another way to see the aura of a person:

Look at a person at their third eye chakra and allow your peripheral vision to see the whole body of the person. While you use a soft focus, you will soon notice the aura of the person. Again, it might start as a mere white color, but you will soon see other colors with enough practice.

The idea here is to be able to switch to your peripheral vision. As you can see, there is nothing magical about this. The reason is that it should be normal to see the aura. It just so happens that the modern world has taught man to think that auric sight is a strange ability when it is, in fact, very simple and common for humans.

Interpreting Aura Colors

Before we leave the topic of aura, let us have a little discussion about interpreting the colors of an aura. Some years ago, an experiment was conducted where different clairvoyants looked at the same person and interpreted their aura. The results seemed disappointing at first. The said clairvoyants saw different colors from each other. Some said that the aura of the person was yellow, while others said that the color was orange, and so on. It seemed as if they were all lying or perhaps just mere charlatans. However, it was soon discovered that the reason why this happened was that auric sight is subjective. The color that you see would depend on what that color means to you. This is why although those clairvoyants saw the aura color differently, they almost all had the same reading with regard to the person's health, state of mind and emotions. Therefore, when you interpret auras, be sure to also pay attention to your own meaning or understanding of the color concerned.

Chapter 15: Vibration

What Is Vibration?

According to Nikola Tesla, if you want to understand the secrets of the universe, then you should think in terms of energy and vibration. You are probably familiar with the meaning of vibration. For example, when you spend a time with a negative person, you might say that the person has a bad vibration. But, when you spend time with a good person, you can say that the person has a good vibration. This is something that many people say without understanding what it really means. Think of vibration as the kind of frequency or station in a radio. If you hit the right frequency, then you can reach a certain mindset. There are vibrations that radiate positive energy, but there are also certain vibrations that you would rather avoid.

In your practice of spirituality, especially if you are just starting out, it is advised that you stick to positive vibrations. Hence, stick to positive people. However, this does not mean that negative people should no longer have a place in your life. After all, it seems unavoidable to have to deal with negative people from time to time. But, do not see these negative people as completely negative. As the saying goes, "Your enemies are the true teachers of virtue." It is the negative people who will teach you the value of patience and self-control. Hence, learn from these negative people. Do not hate or run away from them. Instead, win them over with love and goodness of heart. Remember that energy does not die, but it can change from State to state.

Increase Your Vibration

Just how do you go about increasing your vibration? Well, remember that vibration is just like a frequency. There are high and low frequencies. The important thing to remember is that positive energies have a high frequency while negative energies have lower frequencies. Hence, to increase your vibration, you have to adopt a positive mindset. This is easy. You simply have to focus your mind on positive things, such as love, kindness, peace, understanding, faith, love, and the likes. Take note that this should be accompanied by positive actions. By increasing your vibration, you can do away with all the negative energies in your life.

Now, this is not always an easy thing to do. It is easy to say that you should focus on positive energy, but it is harder to actually do it especially when you are already faced with negative energy. In a difficult situation, you have to exercise some self-control to overcome negative energy with positive energy. Here, you have to make a choice. The key is not to lose your temper and always cling to that which is good.

Acts of kindness and love increase your vibration. Take note that this happens whether you are the giver or the receiver. It is also noteworthy that the practice of meditation is an effective and powerful way to increase vibration. The more that you increase your vibration, the more that it becomes your natural state. Hence, it is recommended that you always strive to increase your vibration. It is also a good defense against negative energy. Take note that positive and negative energies have different vibrations. By increasing your vibration, you get to protect yourself from negativity. Needless to say, if you want to live a good and happy life, then you should work on increasing your vibration since it takes you to a state where you attract energies of love, kindness, and other positive and meaningful energies.

Chapter 16: Mind Power

The Power of Your Mind

Despite the advancement in technology, the mind remains a mystery. There is simply so much about it that is unknown, especially with respect to its vast potentials. As the saying goes, "The All is mind, the universe is mental." If you get to master the mind, you get to master anything. However, it is not that easy to master one's mind. This can easily be proved when you first sit in meditation. It is even hard to keep the mind still and undisturbed by thoughts for a few minutes, especially if you are just a beginner. However, with continuous practice, the mind can be tamed and be under your control. In your spiritual development, it is important that you learn to control the mind as it is the key to all your powers and hidden potentials. It is also with the mind that you influence your chakras and direct prana. To this day and age, the full power of the mind remains unexplored, and the extent of its power remains unknown. This is because there is no limit as to the powers of the mind except the one that you have imposed upon it. The mind is very powerful. Master it.

Expand Your Mind Power

How do you expand the power of your mind? Well, there are no shortcuts for this. The way to do it is by continuous practice. All the practices in this book will help expand your mind power. The question is: which part or attribute of the mind do you want to expand? You see, everything is of the mind. Without the mind, the chakras, energy, and everything else would not make any sense. By now, you already know notable practices that you can use to expand the power of your mind. It is up to you to keep on practicing.

It is also important to note that the power of the mind does not lie in just the mere use of visualization. It also depends on your faith. When you do the exercises in this book, you ought to have faith in what you are doing. Do not just see them as mere visualization exercises. Instead, actually see and feel your chakras and the energy that you imagine. You might want to consider this some kind of sleight of mind if you would. It should be noted that you actually control real energy. This is not just a set of imaginary exercises but would reveal to you the true power of your mind, especially when it is harnessed using visualization and the exercise of the will.

Do not think that you can learn these techniques quickly. The mind usually needs time to adjust. Hence, do not be surprised if you are not able to execute a particular technique despite knowing the right instructions. As we have already talked about, you still need to engage in practice over time. Effective execution is different from knowledge.

It cannot be stressed enough that if you want to change the power of your mind, then you will never go wrong with the practice of meditation. There is one better exercise or training for the mind than a regular practice of meditation. The good news is that every meditation technique is good for the mind so you can have a free choice on which meditation techniques you want to use. Once you have decided on the techniques that best suit you, then all that you need to do is to engage in the actual and continuous practice.

Mind-Healing

Mind-healing or healing with the use of one's mind is another very interesting subject. So, is this real? If you have been keeping a close attention to our discussion, then you should know by now that mind healing is not only real but that it is also something that you can learn and do. In fact, there are techniques in this book that teach you how to heal yourself. Once you learn how to heal yourself, then you can use that ability to heal others.

The power of your mind is only as strong as you make it. In truth, it has no boundaries or limits, but it is only you who usually put a limitation on what it can do. Reiki, pranic healing, and all other healing methods, all depend on the power of the mind.

However, your mind is both a friend and an enemy. If you fail to control it, then it can wreak havoc against you. You must learn to still the mind and be in full control of your thoughts. As the saying goes, "The quality of your life depends upon the quality of your thoughts."

Chapter 17: Enlightenment

How to Achieve Enlightenment

When you take the spiritual path, you will surely encounter the idea of achieving enlightenment. Indeed, many monks and spiritual gurus make this their number one objective. So, how do you receive enlightenment? Well, there is no right or wrong way to attain enlightenment. It may depend on what enlightenment means to you. There are those who say that enlightenment is the way out of the vicious with and rebirth cycle of life. There are also those who think that it is only when you become enlightened that you can truly be free, and there are also those who simply see it as a way to make things for interesting. It does not really matter what it is, for the meaning of enlightenment would depend on the meaning that you give to it. If you are the type who thinks that the only purpose of man is to be enlightened, then perhaps your only goal in your spirituality is nothing but to soon achieve Nirvana or enlightenment. There are also those who make serious sacrifices to earn it, while there are also those who think that enlightenment is beyond the human capacity, regardless of what enlightenment is to you or what you think about it, know that there is always something good to live for in life.

Enlightenment is not about the mere acquisition of powers, but it is a serious step towards a rich and meaningful spiritual life.

Is enlightenment important?

Okay, there are conflicting views on this matter. For some people, enlightenment is the only purpose or goal of their spiritual life; however, for others, achieving enlightenment is not even considered important. In fact, many of them believe that they will not achieve enlightenment in this lifetime. However, this does not mean that they no longer need to do good. After all, you can be full of goodness without achieving enlightenment. When the great Buddha attained enlightenment under the Bodhi tree, he was already a good person even prior to that time. In fact, he had a very rich and strong spiritual life. Of course, after achieving enlightenment, his spiritual life got even richer and more meaningful. So, whether or not enlightenment is important depends on you. If you feel strongly that you need to be enlightened, then you can say that it is important; but, if you are one of the many who does not like to think so much about being enlightened and would rather spend more time getting busy with life, then that is a choice that you can also make.

Chapter 18: Fasting

The Importance of Fasting

Fasting has been in existence for many years. Since ancient times, people have fasted for various reasons, but mostly for spiritual purposes. Fasting has manifold benefits that you can enjoy. It teaches you self-control and discipline. It is also a good way to cleanse both the physical and energy body. There are also people who fast to gain spiritual blessings. As you can see, there are many ways that you can benefit from fasting. Even today, modern science is starting to discover the great health benefits of fasting. Famous spiritual masters are also known to have fasted for long periods. Indeed, there is something about fasting that is richly spiritual.

But, what exactly is fasting? Today, people claim different ways to fast. For example, you can fast by not doing what you enjoy the most. However, the classic definition of fasting is deliberately avoiding food. Hence, it is not eating for a certain period. One of the best ways to fast is known as water fasting whereby you can only drink water during the fasting period. A fast can last for a day to several months. When you fast, it is important for you to listen to your body. Unfortunately, there are people who did not listen to their body and died because of extreme fasting.

Do you really need to fast? Well, probably not, but it cannot be denied that you can gain lots of spiritual benefits from fasting. So, this is something that is strongly recommended that you should try.

How to Do Fasting

Simply put, fasting means not eating. There are also fasts where you also avoid drinking liquids. In a water fast, you are only allowed to drink water during the whole duration of the fast. There is also what is known as dry fasting. This is probably the hardest kind of fasting whereby you do not eat and drink anything, not even water. This is a very powerful kind of fasting, but you should be careful when you follow a dry fast since it can be risky for your body. Once again, you have to learn to listen to your body. If you start to palpitate or when you find your body trembling or any other signal that means that you should end your fast, then you should do it immediately. Although it is good to fast, it is not good to continue it if your body cannot take it anymore.

A good and recommended kind of fasting is known as water fasting. To do this, you only have to drink water for the whole duration of the fast. Do not eat anything or drink anything else except pure, clean water. If you are just starting out, you might want to do a one-day water fast. Once you get used to it, then you might want to try a three-day water fast, and then you can add more days as you feel comfortable.

As mentioned, a dry fast is where you do not eat and drink anything at all, not even water, during the fasting period. Although this is harder than a water fast, it is also known to be more powerful. However, be careful when you apply a dry fast since it can be risky to your health as you might get dehydrated. If you are a beginner, you can just stick to water fasting.

These days, many people are drawn to what is called intermittent fasting or the IF diet. Intermittent fasting has different cycles, but the most common example is where you fast for 16 hours and enjoy an eating window of 8 hours a day. Take note that although this might be good as far as losing weight is concerned, it is not the kind of fasting contemplated in spirituality. In fact, fasting in spirituality has little to do with losing weight. When you fast, your intention matters.

When you fast for a spiritual purpose, you should not focus on your physical body too much. Of course, you can still enjoy the physical benefits of fasting, but do not spend too much time with your physical body. Instead, spend most of your time practicing meditation and doing other things that can help uplift the spirit. Take note that fasting means so much more than not being able to eat for a lengthy period of time, but it is also about how you devote yourself to spirituality that matters.

Chapter 19: Best Practices

To increase your chances of success, here are notable best practices that you should observe:

- Continuous practice

By now, it should be clear to you that continuous practice is very important to your success. Make sure that you engage in the practice of meditation regularly. You do not need to practice all of the techniques in this book, you should at least choose the one that you want to master, and you should devote time and efforts to practice it regularly. A common problem is to procrastinate and let time pass you by without learning or developing anything new. Remember that continuous practice is at the heart of spiritual progress. If you want to awaken the kundalini or progress deeper into spirituality, then you should always make continuous practice a priority.

- Learn from your mistakes

Always learn from your mistakes. No matter how careful you try to be, you will definitely commit mistakes along the way. Do not be hard on yourself. Instead, you should learn from your mistakes. When you commit a mistake, spend some time to reflect on it. Think about how you can avoid committing the same and similar mistakes in the future. You should realize that every mistake is actually a lesson in disguise to help you become a better person. The more that you learn from your mistakes, the more that you can grow and improve.

- Write a journal

Although it is not considered a requirement, many people claim that writing your journal can be very helpful. A journal will allow you to view yourself from a new perspective and allow you to identify your strengths and weaknesses more easily. Do not worry, you do not need to be a professional writer to keep your own journal. However, you do need to update it regularly and be completely honest with everything that you write in your journal.

You can write anything in your journal that is related to your spirituality. Ideally, it should also contain the reasons why you want to take this spiritual path, as well as the lessons and mistakes that you encounter along the way. Your journal should serve as a reflection of yourself in the spiritual path.

Take note that it is not just a tool for writing, but a form of learning when you read and reflect on your writings. Hence, be sure to make time to read the past and current entries in your journal, and then spend some time to reflect on them. Learn as much as you can.

If you are not fond of writing, you might want to just use a file on your computer. These days, mobile phones also come with a free writing application. If you want, you can make use of such an application so you can easily and conveniently write your journal. Just be sure not to lose your file. Of course, you are also free to use the classic pen and notebook style if you prefer.

- Take a break

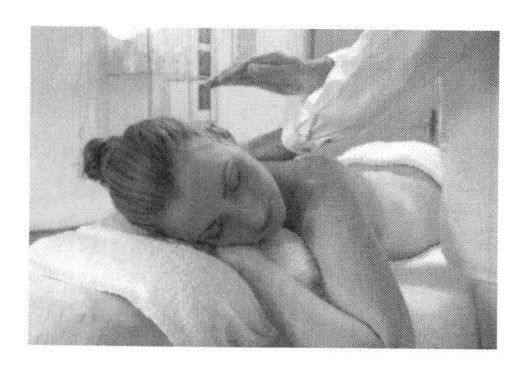

The spiritual path is long and winding one, and you will surely face lots of challenges along the way. The journey would not be fun if you were exhausted. Therefore, you should also give yourself some time to take a break every now and then. Also, when you take a break, do not even think or bother about spiritual matters. Allow yourself to just relax completely and forget about everything else. Do not worry, after this short break, you are expected to train yourself harder.

- Focus on spirituality

It is not uncommon for those who venture into the spiritual path to get lost along the way. The commons cause of this is usually one's greed for power. Indeed, you will develop psychic abilities along the way even if you try to avoid or ignore them. Unfortunately, many people end up thinking that the acquisition of this power is the end to be achieved. But, this is not the case. In fact, such powers are just a part of the spiritual life, so do not be too attached to them. If you allow such power to overcome you, then you will be trapped in its nets, and you will no longer discover the beauty of true spirituality. In fact, even the awaking of the kundalini is just a part of the spiritual journey. It should be important that you learn to control yourself and keep your focus on spiritual growth and maturity.

- Right mindset

If you want to progress in your spiritual life, then you need to think in terms of spirituality. It is time for you to not just see the material value and effects of certain things and actions, but also start considering their effects on the spiritual level. If you have not been making good use of your intuition, then now is the time for you to start paying attention to what it is telling you. You must be open to change and new learnings. If there is no change, then there is no true development. Do not worry, as you ought to change only for the better. As you continue on this journey, you will also get to know yourself better. Be ready to face your inner demons. In case you get confused along the way, just cling to what is good.

In everything that you do, you should give it your best. Unfortunately, there are people who come up with lots of excuses not to give it their best. Normally, this is because they are afraid that their best might not be good enough. Do not be like them. Realize that there is no such thing as a real defeat as long as you keep on trying. However, if you are afraid to even just to do your best, then that is a sign of defeat already. You need to learn to always give it your best and to never stop trying. You have to be strong. True strength is when you change your actions with your whole heart and soul.

- Sacrifice

Sometimes, to progress in your spiritual life, you will have to make sacrifices. For example, instead of always going out with your friends, you may have to skip some meetings and meditate at home instead. Take note that this does not mean that spiritual life should separate you from people. Rather, this only emphasizes that you should not forget about your responsibilities if you also want to make progress on a spiritual level. Just like anything else that is worth having, you may have to make some sacrifices from time to time. Do not worry, the fruit of this journey is very much worth every effort that you put into it. As the saying goes, "The seed is bitter, but the fruit is sweet."

Chapter 20: Road to Mastery

The road to mastery means embracing the teachings and practices in such a way that you make them a way of life. Indeed, this road is also marked by trials and challenges. However, do not let this discourage you. After all, every mountain that is worth climbing always has its own share of risks and obstacles. In fact, if you come to think about it, these obstacles are actually the ones that teach you what you need to learn. Instead of seeing other people's failures as a discouragement, you have to them as a valuable lesson that can help you become a better person.

Now that you are already armed with the right knowledge, I encourage you to choose at least two or three techniques/practices from this book and start practicing them right away. Once again, do not allow failures to discourage you. As long as you keep on doing your best, then you will never lose.

Having knowledge alone is not enough. You need to start right away. The road to mastery is long, so do not waste any more time.

You should also learn how to handle frustrations properly. Even if you possess a strong will, you cannot just advance your spiritual progress as quickly as you want. You will have to spend time learning especially the basics. Also, do not be like the others who compete with other practitioners. Your spiritual path is a path that is for you and you alone. Instead of competing with others, you should draw inspiration from them.

But what does it mean to be a master? A master does not even seek for mastery anymore. Instead, they have turned the techniques into a way of life. A true master is not controlled by power nor do they seek to be more powerful than others. They may have psychic abilities, but they are not a slave to these abilities. As a beginner, you should realize that this spiritual path is one that you should tread with love in your heart.

You cannot be a master if you are enslaved by anger or hatred. In the universe, love remains to be the greatest force. A master knows this little secret and makes sure to always act full of love. Now, this is easy when everything is going your way. But, what if you are met with a difficult situation? Well, this is the time when you will be tested. It is up to you to prove to yourself which side you are really on, whether positive or negative. Just always remember: There is always this great force inside you called love, and you can always choose it above other things, and it is truly more powerful than hatred. Unfortunately, only a few have learned to wield love sincerely.

A word should be said about love. Love is, without a doubt, the greatest force in the whole universe. As you learn the techniques in this book, let love be your guiding light. There is no sense in being able to reach a high level of development if you do not have love in your heart. A common mistake committed by practitioners is becoming too absorbed with themselves. Although it is good to focus on what you are doing, never let it blind you from what is important in life. What good is being able to manipulate all the prana in the world if you do not have people whom you share your love with? Never be blinded by power, and always remember that love should be the foundation of your spirituality.

Conclusion

Thanks for making it through to the end of this book. I hope it was informative and able to provide you with all of the tools you need to achieve your goals whatever they may be.

The next step is to apply everything that you have learned and start practicing the techniques. So, what are you waiting for? Start testing out the techniques in this book and choose the one that you want to master. Always keep in mind that continuous practice is a very important element of success, so be sure to make time and efforts for your practices.

Do not rush the learning process. Instead, you should enjoy it, and learn as much as you can from them. Indeed, this book is more than a collection of texts that reveal ancient secrets and wisdom. Rather, this book is an invitation to a life-changing journey. May this book be a guide like a shining star that leads you to the true path of spirituality. Just the fact that you have read this book only means that you have some kind of spiritual hunger inside you. Once again, knowledge and actual application are necessary to satisfy and make progress in your spiritual life. Feel free to review the pages of this book and be sure that you understand all of the teachings and instructions. With regard to the practices, you are allowed to make some adjustments or changes according to your preference as you deem best and suitable for you. Always remember to always do your best and never give up.

Indeed, so many people have been searching for the right instructions that can help them awaken the kundalini, develop psychic abilities, or even at least to make some progress in their spiritual life, but in vain. Now, this book has given you the keys to power, happiness, and a beautiful life. It is up to you to put this new-found knowledge into actual and continuous practice. Last but not least, remember to use everything that you have learned from this book only for good.

I am glad that you have reached this part of the book. I really hope that you enjoyed the read. If you truly desire to advance in your spiritual life, then know that it is possible, but you need to put in the time and efforts to make it a reality. You can do it as long as you continue to do your best.

Finally, if you found this book useful in any way, a review on Amazon is always appreciated!

Made in the USA
San Bernardino, CA
27 June 2019